Adventures
in the
Science
of
Cosmology

Ken Dixon

To my only grandchild,
Parker Kenneth Lee-Dixon,
who may (or may not!) decide to read this
when he reaches the age of thirteen.

Contents

List of Figures

List of Tables

List of Key Formulas

Preface

To all you readers curious enough to open this book, I thank you. I hope you will find it both informative and entertaining — perhaps even a refreshing break from your everyday busy life, whatever your age or background. If you like science but have never given astronomy or cosmology much thought, this book is definitely for you ... but only if you fall within the age range of 13 to 113.

Having said that, however, my focus is on young teens. The goal here is to introduce young minds to the scientific and philosophical wonders of astronomy, with an emphasis on cosmology. The book is intended for students

- who are in grade nine or higher (around 13 to 16 years of age)
- who have had some math and science, but perhaps not much
- and — most importantly — who have an aptitude for science (whether or not they fully realize it at this stage of their schooling).

But whatever your age, if you would like to learn more about the workings of the universe — and in particular, how astronomers use science and math to figure it all out — then welcome aboard. May you find fact stranger than fiction within these pages.

Part I presents lots of facts about the universe, including some weird ones that may surprise you. It introduces some of the incredibly large numbers that astronomers deal with. You will learn, for example, that the Andromeda galaxy contains roughly one trillion stars and is about 2.5 million light-years away. But it is moving towards our own Milky Way galaxy at about 120 kilometers *every single second* (that's about 270,000 miles per hour). If this worries you, you might want to pull out your scientific calculator (you do have one, right?) and figure out roughly when this mother of all collisions might occur — assuming, of course, you already know what a light-year is.

Part II is the heart of the book. It explains in simple steps how astronomers use science and math to determine some of the facts and figures presented in Part I. You will learn, for example, how they figure out a galaxy's distance from us and the speed it is moving toward or away from us. (Hint: A really long tape measure and one of those speed trap radar guns won't cut it.) You will learn how they calculate the age of the universe. You will learn where all the chemical elements come from. In short, you'll learn something about how the universe works — or at least how astronomers, cosmologists, and particle physicists *think* it works. I suppose they could all be wrong, but I doubt it somehow.

Part III is a sober reminder that there are still lots of fundamentally deep, unsolved mysteries for the next generation of scientists to tackle. For example, these geeky, wiz-bang men and women only understand four percent of today's universe. Yep, you read it right — *only four percent*!!! The other 96 percent is made up of what they call "dark matter" and "dark energy", but nobody knows what this stuff really is. How cool is that? This situation is downright embarrassing, so it's always a fun topic to bring up at any party where cosmologists are present. (You do go to such parties, right?)

Who knows, perhaps one day, you will help solve some of these mysteries. Yes, *YOU*. After all, why should someone else have all the fun and glory?

Finally (sigh) … I suppose I should introduce my good buddy and self-appointed writing assistant, Tri-face:

He pops up from time to time, wearing one of these three faces. Some of his comments are pretty silly and he likes to laugh at his own jokes. Feel free to ignore him whenever you wish. I often do. Personally, I find him annoying at times, especially when he brings up awkward points that I then have to try to explain to you readers.

On the other hand, some of Tri-face's comments are quite informative, if not downright philosophical. His three different faces usually give a clue as to what he's up to. But not always.

Ken Dixon
Toronto, Ontario
May, 2013
kdixon164@rogers.com

Part I

A Quick Look
at the Universe

Chapter 1 —
Really Large Stuff

Before we start, we need to explain a few key words. By **universe** or **cosmos** we mean all the matter and energy that exists throughout space, in whatever form. And that includes space itself. In other words everything, including you and all your pals. Even your teachers.

By **astronomy** we mean the study of all the objects in the night sky that lie beyond Earth's atmosphere, whether visible or not (you can't actually see a black hole). Such objects are often referred to as **celestial objects**. By **cosmology** we mean the study of the origin, evolution, and large-scale structure of the universe.

Think of cosmologists as geeky scientists who study the big picture — the whole universe.

So what makes up the matter in our universe? Or, to put it another way, what is the structure of the cosmos? This is probably not the sort of question that comes up a whole lot when you're playing video games, texting friends, or talking on the cell phone. Or, for that matter, doing homework.

Let's face it, we can't all be curious about everything all the time. But when we do think about this particular question, we tend to picture what we can see with the naked eye or with the aid of a telescope. We see all the stuff around us in our everyday world, plus all the really big stuff beyond. Like stars and planets.

But if we want to understand the science of astronomy and cosmology, we need to recognize that the universe also consists of really small stuff that we can't see with the naked eye. We can't see tiny things like atoms and subatomic particles that actually make up the big stuff. Most of these things are so small you can't even see them with a powerful electronic microscope.

We'll call the big stuff the **macro world** and the small stuff the **micro world**. We need a knowledge of both if we are to understand the history and evolution of the universe from the time it began. (And yes, according to the Big Bang theory, it did have a beginning.) In this chapter, we'll look at the big stuff and leave the small stuff for Chapter 2.

The Macro World

Figure 1-1 is a simplified picture of the macro world. The objects included here are very large compared to what you're used to dealing with, like baseballs, dogs, people, buildings, and mountains. But most of them are so incredibly far away that you need a telescope to see them. Let's have a brief look, starting with our own astronomical neighborhood and working out from there.

Our Solar System

This component is very close to us — after all, we're part of it! Earth is the third **inner planet** from the sun, Mercury being the closest, Venus the second closest, and Mars the fourth. These four bodies are also known as the **terrestrial planets** because that word relates to land, as opposed to gas or liquid. The whole surface of these four planets consists of solid land, with the exception of Earth where there is actually more ocean than land.

 Did you know that if you laid all the world's astronomers end to end around the globe, two thirds of them would drown?

I think Tri-face is trying to tell us that Earth's surface is roughly 33 percent land and 67 percent water. Good thing, too. Without water, it's unlikely life would exist.

Beyond Mars are many thousands of asteroids that form the **asteroid belt** between Mars and Jupiter (the first of the outer planets). These rocky bodies range in size from mere boulders to a few hundred kilometers. Almost all of them lack enough mass to create an internal gravitational pull strong enough to compress them into a ball. So their shapes are weird and irregular, like the potato-shaped asteroid Ida in Figure 1-2. Ida is about 38 kilometers long and 22 wide.

Ida was photographed by the robot spacecraft Galileo in 1993 on its way to Jupiter. It was the first discovery of an asteroid having a moon. That's the tiny dot to the right of Ida and it's only 1.5 kilometers across. Many asteroids are now known to have moons.

 I see our illustrious author has mentioned kilometers. The scientific community uses the metric system in which the units come in multiples of ten. Really simple. (See the Reference Data section in the back of the book.) If you are more familiar with that other system where one mile equals 5,280 feet (ugh), just remember that one kilometer is equal to 0.62 miles. Think of it as a short mile. So ... Ida's moon is about one mile across.

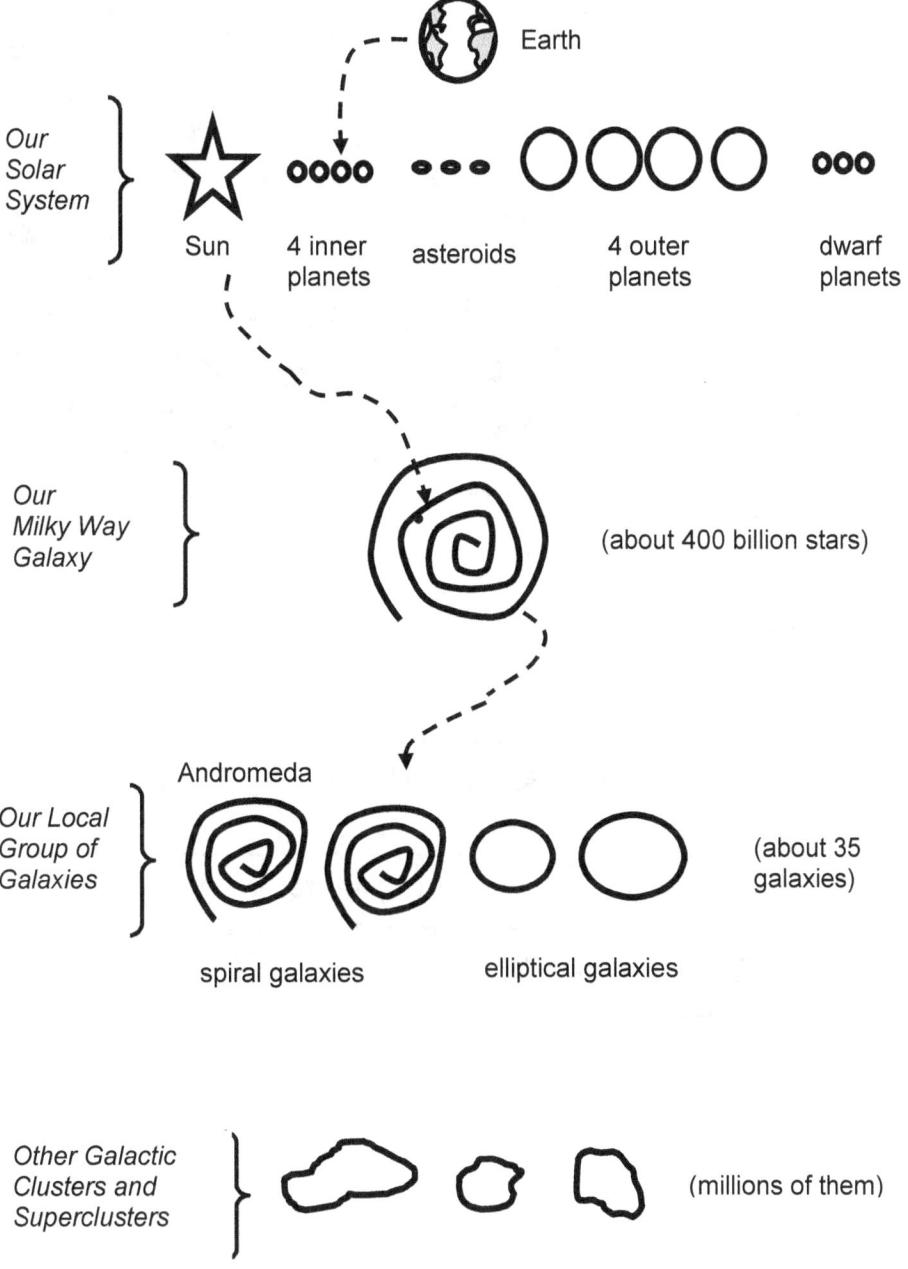

Figure 1-1 Macro World

Beyond the asteroid belt lie the four large **outer planets**, also known as the **giant gas planets** — Jupiter, Saturn, Uranus, and Neptune in that order. Unlike the terrestrial planets, these giant planets have a surface of gas, mostly hydrogen. Not an ideal place for walking or hiking.

Figure 1-2 Asteroid Ida and Its Moon

Beyond the outer planets is a large number of small bodies of ice and rock that form what is called the **Kuiper Belt**. These bodies are known as the **dwarf planets**. Only a few of them have been identified and named. Pluto, of course, is the most famous one, discovered in 1930 and considered to be the ninth planet. That's until 2006 when astronomers realized that there were probably a great number of Pluto-like bodies to be discovered in this outer region and that therefore Pluto did not deserve full planet status.

 Rumor has it that the Plutoites living there are really mad about this and have voted to secede from the solar system.

Yeah, right.

So what's missing from the top part of Figure 1-1? Well, we didn't show the **satellites** (moons) that orbit some of the planets, nor the icy comets that go whizzing past Earth in spectacular fashion from time to time in their elongated orbits around the sun.

All the solar system objects move around the sun in their own orbits with their own length of year. These orbits, however, do not form perfect circles. Each one is somewhat egg shaped, a mathematically defined shape known as an **ellipse**, in which the distance between the sun and the object varies during its annual orbit.

For our planet, the *average* Earth-to-sun distance is about 150 million kilometers. Astronomers call this distance an Astronomical Unit (AU). They often use it as a yard stick for measuring distances within the solar system. For example, Jupiter is 5.2 AU from the sun. Because of its slightly elliptical orbit, Earth actually varies from 0.98 to 1.02 AU in its distance from the sun during the year.

The sun has almost a thousand times more mass than all the planets, moons, asteroids, and comets combined. And — as everybody knows — the sun is a star.

Finally, most celestial objects spin like a top. That is, they rotate on their axis. Earth's rotational period is 23 hours and 56.1 minutes. The sun's rotational period is roughly four weeks.

Our Milky Way Galaxy

Our sun is just one of roughly 400 billion stars that make up the Milky Way galaxy. This number varies, depending on who you talk to. Some things in astronomy are hard to pin down with a lot of certainty.

Imagine traveling far enough in space to view our galaxy from above, and then edge on. We would see it as a flattened, spiral shaped disk with a bulge in the center and large spirals of stars swarming around this bulge. Figure 1-1 shows our sun located along one of these spiral arms. It is a long way from the galactic center but well within the outer edge of the galactic plane.

If you jump ahead to Figure 1-3, you'll see what a spiral galaxy looks like when viewed at some angle between directly above and edge on. Like stars and planets, galaxies also rotate on their axis. It takes our sun 220 million years to complete one trip around the galaxy. This is not your typical merry-go-round.

Because most of the stars in our galaxy are located within the plane of the disk, we see a dense "Milky Way" band of stars when we look into the night sky toward the galactic center. That's because we are looking along (or through) the plane of the disk rather than up or down from it (perpendicular to it) where there are relatively few stars.

So what holds these billions of stars together as a galaxy? Why don't they just fly off in all directions and go their own way? The answer is the same one that holds our solar system together — gravity. Astronomers say that these systems, whether a solar system or a galaxy, are *gravitationally bound*. The force of gravity, created by the mass within the system, holds all that mass together. It is very difficult for a star to escape the gravitational pull of the galaxy, just as it is very difficult for an object on Earth to escape the gravitational pull of our planet. Have you tried jumping off lately?

Every astronomical body (star, planet, and so on) has its own **escape velocity** — the vertical velocity which an object on its surface must reach to overcome the gravitational pull of that body and escape into space. Earth's escape velocity is 11.2

kilometers per second. The escape velocity of the sun (a much more massive object) is 617.4 kilometers per second.

Our Local Group of Galaxies

Galaxies tend to cluster together and the third part of Figure 1-1 tells us that our Milky Way galaxy belongs to a local group of roughly 35 galaxies. Most of them are much smaller than our galaxy or our sister galaxy, Andromeda, which is largest one in the group (see Figure 1-3). Some of the smaller galaxies that are very close to our

Figure 1-3 Andromeda Galaxy (roughly one trillion stars)

galaxy are actually **satellites** of our galaxy — they revolve around the Milky Way galaxy just like the moon revolves around Earth. Some of the local galaxies are quite different from spiral galaxies. They are more egg shaped or spherical than pancake shaped, which is why they are called **elliptical galaxies**.

So what keeps these galaxies together as a group? They are gravitationally bound. They are very close to each other compared to the much greater distances that lie between this cluster and neighboring clusters.

Other Galactic Clusters and Superclusters

The bottom part of Figure 1-1 tells us that there are many other galactic clusters beyond our local group and that throughout the vastness of the universe there are actually groups of clusters called superclusters. And — as everybody knows — the clusters of a supercluster hang out together because they are gravitationally bound.

Enough Different Motions Here to Make You Dizzy

Of course, none of these cosmic components stand still. Everything is in constant motion. Have you ever wondered how many astronomical motions you take part in — all at the same time — while standing, say, on the equator of our planet? You don't actually feel these motions because you're sharing them with the surrounding land and air. The land and air are moving right along with you. Here are the major motions (yes, there are lesser ones as well, but we'll ignore them here):

1. You are rotating with Earth around its axis about 1,670 kilometers per hour (abbreviated 1,670 km/h) at the equator. That's about 17 times faster than the family car goes. Or *should* go!

2. You are orbiting with Earth around the sun about 30 kilometers per second (abbreviated 30 km/s). That's about 1,000 times faster than the family car.

3. You are orbiting with the sun around the center of our galaxy (because the sun is dragging Earth along with it) about 220 km/s. That's about 8,000 times faster than the family car.

4. You are moving with our galaxy within the local group of galaxies (because our galaxy is dragging the sun along with it). For example, it is estimated that our galaxy and the Andromeda galaxy are heading toward each other at a relative speed of 120 km/s. (Yes, we may collide some day, as discussed in Chapter 7.)

5. You are moving with the local group of galaxies (because it is dragging our galaxy along with it) towards other galactic clusters at various relative speeds.

6. You are also moving along *with space itself* in the expansion of the universe. But that's a totally different kind of "motion", best left for a later chapter. Unlike the other motions, this last one is not motion *through* space.

Although you don't actually feel these motions, you can experience them indirectly by observing the effects they have on the position of celestial objects in the sky. You know that during 24 hours you can see big changes in the sky due to motion 1 (it's like night and day!). And during 365¼ days, you can see other changes in the sky due to motion 2 (like the position of the sun and the constellations). If you could live for another 220 million years, you could see more changes in the sky due to motion 3, like the position of the stars and maybe even a bit of shifting of nearby galaxies due to motion 4. No need to go on. You get the idea.

Newton's Universal Law of Gravitation

As you probably know from any physics you've taken at school, things don't move around unless a force is applied to them. Most of the celestial motions we have mentioned here involve the force of gravity. According to Isaac Newton back in the 1600s, any two objects in the universe are attracted toward each other by gravity, no matter how far apart they are. They may or may not eventually collide, depending on what other bodies, forces, or velocities are involved in the dynamics.

Here is a formal statement of Newton's famous universal law of gravitation:

> Two bodies attract each other with a gravitational force that is directly proportional to the mass of each body and inversely proportional to the square of the distance between them.

Newton considered gravity to be a mysterious force field that acts instantaneously across the distance between objects, but he had no idea how it actually worked. What was the mechanism? He could only measure its effect in the laboratory and by observing the motion of planets (not to mention falling apples).

NOTE: A law like this is considered to be a **theory**. This law comes up again in Chapter 8 when we discuss what scientists mean by the word "theory". Like most theories in physics, this one can be expressed as a mathematical formula (or equation).

Mathematically, Newton's law looks like this, where the horizontal line means divided by:

$$F = \frac{Gm_1m_2}{d^2}$$

where:

F = gravitational force between two objects
G = universal constant of gravity
m_1 = mass of first object
m_2 = mass of second object
d = distance between them.

So what does this equation mean? Well, it means that an increase in mass of either object would increase the force between them while an increase in distance would decrease that force. Why decrease? Because the force between the objects gets *divided* by the distance. More specifically, if you double the mass of either object, the force will double. If you double the distance, the force will be reduced to a *quarter* of what it was (not a half, because distance is *squared* in the formula).

 This might be a good place to hold a brief lesson on equations — like, for anyone who may not have messed around with algebra yet.

Good idea, Tri-face. We don't want anyone suddenly coming down with an acute, algebraic, anxiety attack. They'll need the lesson for Part II anyway. It's hard to explain certain scientific concepts without using algebra. After all, mathematics is the language of science. Without math, science would be crippled.

So here we go — A REALLY BRIEF LESSON ON EQUATIONS.

The mathematical relationship presented above is an **algebraic equation** or **formula**, where the stuff left of the equal sign (=) is equal to the stuff right of the equal sign. Think of an equation as a balancing scale with equal weights on each side as shown in Figure 1-4. Whatever you do with this equation, you must keep it in balance.

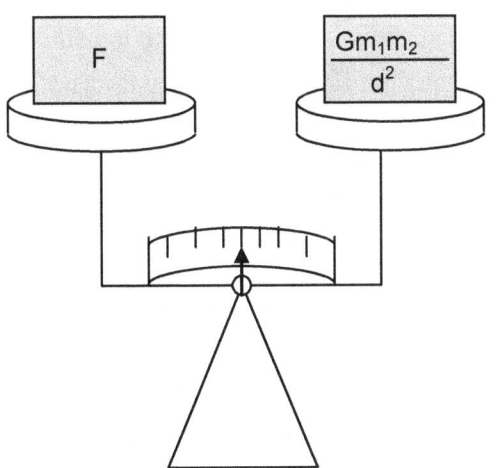

Figure 1-4 Newton's Law

The symbols F, G, m_1, m_2, and d are the individual **terms** that make up the equation. If two terms are to be added together, a plus sign (+) is shown between them. If one term is to be subtracted from another, a minus sign (-) is shown between them. Neither situation occurs in the case of Newton's law.

When one term (or group of terms) is to be divided by another, a horizontal line is shown with the quantity to be divided placed above the line and the quantity doing the dividing placed below it. We don't use the division sign (÷) in this book. When two terms are to be multiplied together, normally no sign is shown between them. A multiplication sign (x) is assumed. So Gm_1m_2 actually means:

$$G \times m_1 \times m_2$$

Newton's law shows both division and multiplication.

Except for G, all the terms in Newton's law are called **variables** because their values can vary, depending on the particular situation represented by the equation. G is a **constant**, meaning it has only one value and it never changes. In this case, it's one of those physical constants that make up the laws of nature (like the speed of light discussed in Chapter 9). The Reference Data section in the back of the book gives the value of a few of these physical constants.

Now, if you know the value of all the variables except one, you can calculate the value of that **unknown variable**. In algebra that's referred to as *solving for an unknown*. So if someone asked you to determine the gravitational force (F) between Earth and the sun, you could go to a textbook, or a website, or the back of this book to look up the mass of these two objects (m_1 and m_2) and the distance (d) between them. Oh yeah, and grab the value of that constant G while you're at it. Then you would have everything you need to calculate F. Except maybe a scientific calculator to handle those big numbers.

Sometimes you have to rearrange the equation in order to solve for the unknown. For example, if I gave you the value of F, m_1, and d, and asked you to solve for m_2, you would have to do some rearranging of the terms to get m_2 on one side of the equation *all by itself*. Then you could calculate it from all the known values on the other side. The secret to rearranging an equation is to remember that the balance scale must always be kept in balance. This means ...

... *whatever you do on one side of the equation, you must also do on the other.*

In the above problem, you can get m_2 on the right side all by itself by dividing that side by Gm_1 (to cancel out the Gm_1 on the top) and multiplying it by d^2 (to cancel out the d^2 on the bottom). So far, so good. Now to keep the equation in balance, you have to do the same thing on the left side — divide by Gm_1 and multiply by d^2. The result:

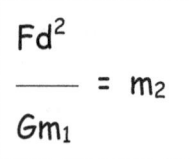

$$\frac{Fd^2}{Gm_1} = m_2$$

Now you can calculate m_2 from the known values. We'll do a lot of this rearranging in Part II. It's no big deal — END OF LESSON.

We won't be working with Newton's equation in this book, so don't worry about what units to use for force, mass, and distance. But astronomers make great use of it when dealing with such things as planetary orbits and the rotation of galaxies. Falling apples, not so much.

Stars and Sand and Carl Sagan

Here's a final question about the macro world — how many stars are there in the whole universe? Certainly too many to count, even if we could see them all. But we can't even see all the stars in our own galaxy. And the universe is too big to see all the galaxies. Astronomers can only make rough estimates, based on what they know about galactic structures and the distribution and sizes of galaxies. Observations with the Hubble Space Telescope over a small area of the sky provide this information.

For our purpose here, we are going to use two estimates and one key assumption that are typical of recent studies. The first estimate is 400 billion stars in our Milky Way galaxy and the second estimate is 125 billion galaxies in the universe. The key assumption is that our galaxy is an average galaxy with respect to number of stars — not an unreasonable assumption when you look at the size of thousands of other galaxies.

So if you multiply these two numbers together, you should get 50,000 billion billion for the total number of stars in the universe.

That's ... let's see now ... 50,000,000,000,000,000,000,000 stars.

Cool. That's what I call zillions. Now, to imagine how many stars that is, all you have to do is picture what a billion billion looks like and then simply imagine 50,000 of those. Right?

Yeah, right.

Maybe we should just rewrite this ridiculously huge number the way scientists do by using powers of ten. That's a 10 with a tiny number on top of it (called an **exponent**). It tells us how many times we have to multiply the 10 by itself. Or to put it another way, it's the number of places we would have to move the decimal point to the right if we didn't use powers of ten. For example:

$1.0 \times 10^3 = 1,000$ (a thousand)

$1.0 \times 10^6 = 1,000,000$ (a million)

$1.0 \times 10^9 = 1,000,000,000$ (a billion)

$1.0 \times 10^{12} = 1,000,000,000,000$ (a trillion)

$7.82 \times 10^3 = 7,820$

So, if we change each of the three numbers given in the "50,000 billion billion" answer to a power of ten, we get:

$$50,000 \text{ billion billion} = (5 \times 10^4) \times (10^9) \times (10^9)$$

Adding those powers of ten together gives us the number of stars as:

$$50,000 \text{ billion billion} = 5 \times 10^4 \times 10^9 \times 10^9 = 5 \times 10^{22}$$

Now, you gotta admit that 5×10^{22} is a little neater than writing 50,000,000,000,000,000,000,000.

But if you want to visualize how big this number is, try comparing it with some other huge number. Like, say, grains of sand. Picture yourself walking on a sandy beach that stretches a long distance along the shoreline. It could be in Florida,

Hawaii, the north shore of Prince Edward Island, the Mediterranean, or the east coast of Australia. Any place that has a beach, big or small.

Now bend down and scoop up a handful of dry sand and look at all those tiny grains you're holding. Open your hand and watch them as they slide between your fingers and fall back onto the beach.

How many grains of sand do you figure you had in your hand? Hundreds? Probably more like thousands and thousands. Now lift your eyes and scan as much of the beach as you can see. How many grains do you figure there are on that one beach? Hard to imagine, right? As Tri-face would say, zillions.

Now, think long and hard about what the American astronomer Carl Sagan said in his famous TV series *Cosmos* back in the 1970s:

"*The total number of stars in the universe is greater than all the grains of sand on all the beaches on planet Earth.*"

That's pretty breathtaking. And, as we shall see in Chapter 3, there are incredible distances between one star and the next — making it all that much harder to comprehend just how big the universe must be.

Chapter 2 —
Really Small Stuff

In terms of size, think of us humans as being smack-dab in the middle of the macro and micro worlds. Just to illustrate, let's compare the size of a six foot person to (a) the sun and (b) a typical atom. Assuming the person is a scientist, he or she would probably say they were 1.8 meters tall (abbreviated 1.8 m).

The diameter of our sun is 1,392,000 km, so if you multiply that by 1,000 to convert it to meters and then divide by 1.8 m to compare this size with our scientist, you get 773,333,333.3 or (in keeping with the accuracy of the input figures) 7.7×10^8.

"So what?", I hear you mutter. Well, that number is close to a billion (10^9), so something in the macro world like the sun is roughly a billion times bigger than an adult human.

The diameter of a typical atom is around 0.0000001 centimeters (abbreviated 0.0000001 cm), so if you divide that by 100 to convert it to meters and then divide by 1.8 m to compare this size with our scientist, you get 0.0000000005555555555 or (in keeping with the accuracy of the input figures) 5.6×10^{-10}.

Numbers much smaller than 1 are expressed with a *negative* power of ten which tells us how many places we would have to move the decimal point to the *left* if we didn't use powers of ten.

This number is between one billionth and a tenth of one billionth, so something in the micro world like an atom is roughly a billion times smaller than an adult human. So there we are — smack-dab in the middle of the macro and micro worlds.

NOTE: I don't know about you, but I find it easier to picture the enormous size of the sun than the ridiculously small size of an atom. Based on the atomic size given above, it would take roughly 20 billion atoms just to cover the period at the end of this sentence. But don't try this at home.

The Micro World

Figure 2-1 is a simplified picture of the micro world, the basic components of matter that make up the macro world. These components appeared in the very early universe, long before any stars appeared, as we will see in Chapter 12. They are too small to be seen with the naked eye. Some powerful electron microscopes can bring single atoms into view. Anything smaller can only be detected indirectly by reactions in high speed particle accelerators or deep underground particle detectors. But that's another subject.

Let's have a look at these components, starting with atoms and molecules and working down to subatomic particles.

Matter

The stuff we encounter everyday — including our own bodies — is made up of atoms and combinations of atoms known as molecules. A molecule is simply a chemical bonding of two or more atoms. The oxygen in the air, for example, is in the form of molecules made up of two oxygen atoms. The symbol for oxygen is O, so the molecule is symbolized as O_2. Water is a molecule made up of two hydrogen (H) atoms and one oxygen atom, so it is symbolized as H_2O. But you knew that.

We have already illustrated how small the atom is by comparing it with a period on the page. But molecules are still pretty small too. A teaspoon of water contains 1.7×10^{23} H_2O molecules. That's slightly more than the estimated number of stars in the universe that we made at the end of Chapter 1.

The ancient Greek philosophers had the right idea when they reasoned that all matter must be made up of some basic particle that can't be further subdivided. They named this unseen particle the **atom**, which meant "uncuttable". Modern science has retained the name, but the atom is no longer considered uncuttable. It has its own components.

Atoms

As Figure 2-1 shows, an atom consists of an electrically charged nucleus surrounded by one or more subatomic particles called **electrons**. What the diagram does not show is that these electrons are whirling around the nucleus at different distances from it and at speeds in the hundreds, even thousands, of kilometers per second. Try to imagine this kind of dynamic activity going on inside every atom of your body every moment!

The nucleus has an electrical charge of one or more positive units, while each electron has a negative charge of one unit. Normally, the atom is electrically neutral, having the same number of electrons as the positive charge of the nucleus. It is this positive charge that determines what kind of atom it is, for example — hydrogen, helium, or carbon. The different kinds of atoms are, in fact, the chemical elements that make up matter. We'll talk more about this later in the chapter.

Now if you want a meaningful cosmic number that is a lot larger than the number of stars in the universe (5×10^{22}), it would be the total number of atoms in the universe, estimated to be roughly 10^{78}. And — as everybody knows — that's a 1 followed by 78 zeroes. That's too many zeroes to fit on one line of this page.

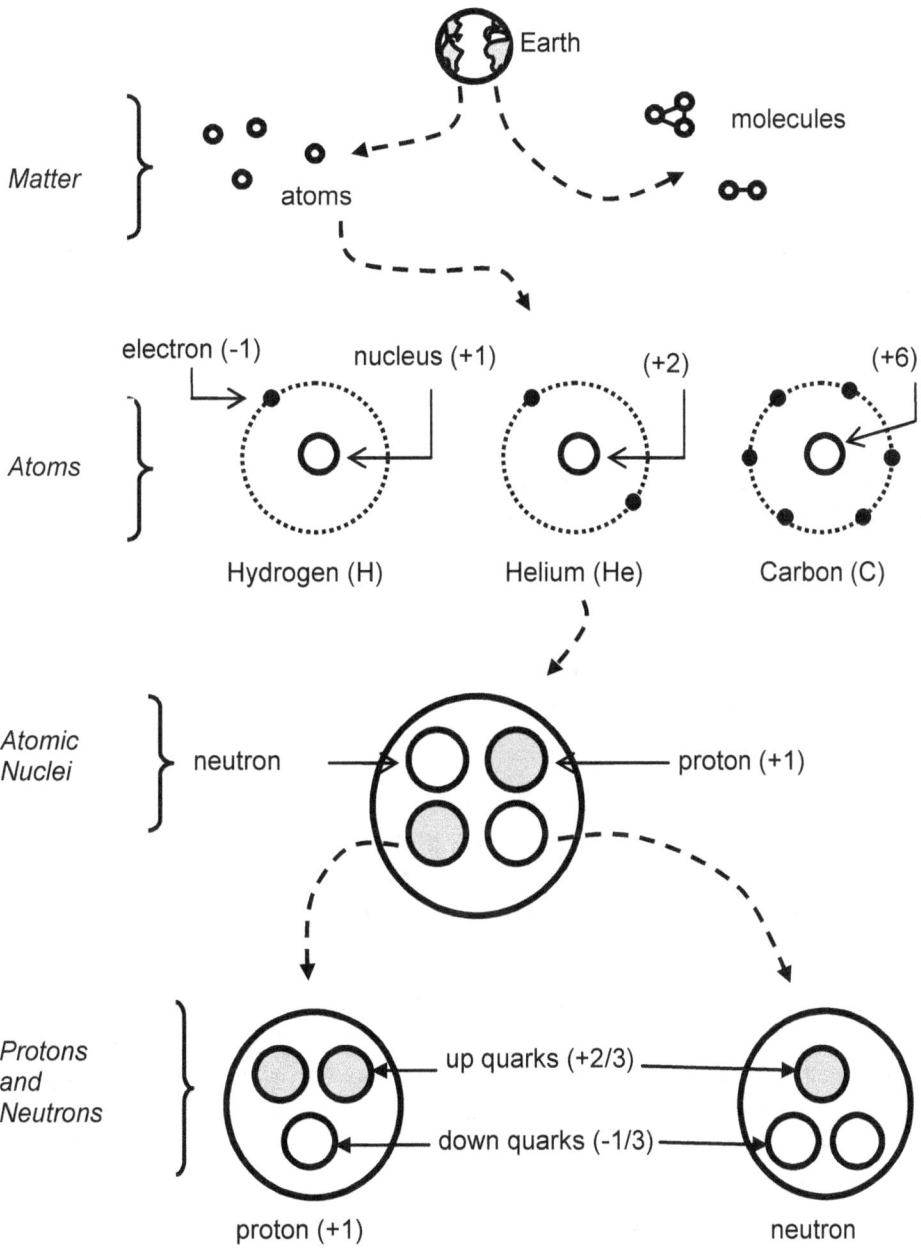

Figure 2-1 Micro World

Atomic Nuclei

The nucleus is made up of two kinds of subatomic particles — **protons**, each with a positive charge of one, and **neutrons**, with no electrical charge.

That's why they call 'em neutrons. They're electrically neutral.

So it is actually the number of protons that determines what kind of atom we have. For example, hydrogen atoms have one proton, helium atoms two, and carbon atoms six. The more protons a nucleus has, the more neutrons it has, but it doesn't have to be the same number of each. The mass of a proton or neutron is roughly 2,000 times more than that of an electron, so it's mostly the number of protons and neutrons that determine the mass of an atom.

A particle of matter that is not made up of smaller components is called an **elementary particle** by the particle physicists. These are the folks who make a living by thinking about the micro world. They are actually a little weird, just like astronomers. They make up strange names for little-known subatomic particles and have developed a whole theory on the micro world called the **Standard Model**. They tell us that the electron is an elementary particle. Protons and neutrons, however, are not elementary particles.

Protons and Neutrons

Protons and neutrons are made up of three smaller particles called **quarks**. And that ends the subdividing. Quarks are elementary particles (at least, according to the current Standard Model). Really small ones. If it takes 20 billion atoms to cover a period on this page, just think how many zillions and zillions of quarks it would take to do the same thing.

As Figure 2-1 shows, there are two kinds (or two "flavors" as these theoretical physicists like to say) of quarks in protons and neutrons — **up quarks** (with a positive electrical charge of two thirds) and **down quarks** (with a negative charge of one third). Notice that the proton has two up quarks and one down quark. If you add up their electrical charges (⅔ + ⅔ - ⅓), you get a total of one — the charge of a proton. The neutron, however, has one up and two downs (⅔ - ⅓ - ⅓), giving a total charge of zero — the charge of a neutron. Pretty cool, eh?

Matter Particles and Force Particles

We have mentioned three elementary particles so far — electrons, up quarks, and down quarks. There are several others that make up matter. There is also a different set of particles that transmit the fundamental forces that act between matter particles, but this is quite another subject. Right now, out of all these other particles, we will only mention two, since they will come up later in the book: **neutrinos** and **photons**.

The neutrino is an elementary *matter* particle that comes in three different flavors, all of which have no electrical charge and almost no mass. In comparison, the tiny electron is a massive giant. Neutrinos travel at almost the speed of light. The photon is a *force* particle that makes up light and also binds electrons and nuclei into atoms via the electromagnetic force. It has no electrical charge, no mass, and travels at the speed of light.

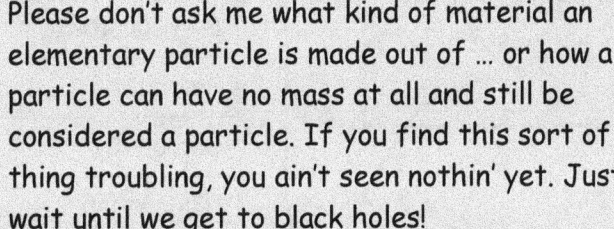

Please don't ask me what kind of material an elementary particle is made out of ... or how a particle can have no mass at all and still be considered a particle. If you find this sort of thing troubling, you ain't seen nothin' yet. Just wait until we get to black holes!

The Periodic Table

We have mentioned hydrogen, helium, and carbon as three kinds of atoms or chemical elements. So how many different kinds are there in the universe? The answer is 92 naturally occurring elements. They start with hydrogen as the lightest with only one proton, then helium the next lightest with two protons. Over the years scientists have established the existence of all 92 elements and arranged them in a **periodic table** from the lightest to the heaviest. As you move through the table, each heavier element has one more proton than the previous one. The number of protons an element has is known as its **atomic number**.

Table 2-1 lists just a few of the more well known elements. Uranium is the heaviest with an atomic number of 92. As Tri-face points out in the table, astronomers use some weird terminology. They don't group these elements the way other scientists do. Astronomers talk a lot about hydrogen and helium, but usually lump the other 90 elements together by simply referring to them as "metals". They do this even though many of them (like carbon and oxygen) are obviously not metals in the normal sense of the word.

So why do astronomers do this? Because they have a very different perspective on the matter (no pun intended). As we will learn in Part II, hydrogen and helium came into existence in quite a different way than the other elements and make up roughly 98 percent of the ordinary matter in the universe. All the other elements combined (those so-called "metals") make up the remaining 2 percent.

Astronomers are an odd bunch. They refer to all the elements except H and He as "metals".

Chemical Element (and its symbol)	Atomic Number
Hydrogen (H)	1
Helium (He)	2
Lithium (Li)	3
Beryllium (Be)	4
Boron (B)	5
Carbon (C)	6
Nitrogen (N)	7
Oxygen (O)	8
Iron (Fe)	26
Silver (Ag)	47
Gold (Au)	79
Lead (Pb)	82
Uranium (U)	92

Table 2-1 A Few Elements from the Periodic Table

NOTE: Although helium is, by far, the second most plentiful element in the universe, it is the only element that was first discovered somewhere other than Earth. Aside from the odd birthday balloon floating around, helium is actually quite rare on Earth. It was discovered in the spectrum of the sun which is composed of 25 percent helium by mass. Chapter 9 explains how scientists identify elements in the sun.

Quantum Mechanics

The Standard Model of particle physics includes a concept called **quantum mechanics**. The value of a property of a particle (such as its electric charge) comes in specific amounts. The same applies to the energy level of electrons within an atom. These values do not cover a continuous range. They can only change in small discrete amounts known as **quanta** (the plural of quantum).

Another example of this quantum principle lies in the nature of light (as you will see in Chapter 9). A beam of light is made up of tiny packets or quanta called photons. You can't break light down into fractions of a photon or quantum.

I want to mention one quantum characteristic of electrons that turns out to be extremely useful to astronomers in their study of the macro world. I didn't bother showing this characteristic for the carbon atom in Figure 2-1, but the fact is that electrons whiz around the nucleus of the atom in different orbits.

Today's particle physicists consider these orbits as *different energy levels*. It seems mother nature has laid down the following laws for electrons:

1. Each orbit represents a very specific energy level for the electrons in that orbit.
2. No electron can have a energy level that lies between two adjacent orbits. They can only move from one orbit to another.
3. At the innermost orbit (lowest energy level) there is only room for two electrons.
4. At the second energy level there is room for eight electrons.
5. At the third energy level there is room for 18 electrons.
6. At the fourth energy level there is room for 32 electrons.
7. And so on for higher energy levels in the electron structure of heavier elements.
8. If there is room for an electron to move from a higher energy level to a lower one, it will try to do so to keep the whole atom at the lowest energy level possible.

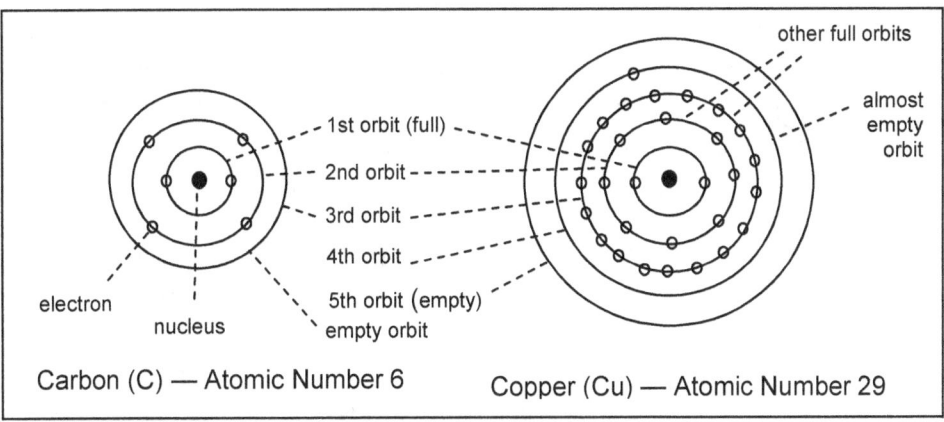

Carbon (C) — Atomic Number 6   Copper (Cu) — Atomic Number 29

Figure 2-2 Electron Configuration in Carbon and Copper Atoms

Figure 2-2 contains a more complete picture of the carbon atom with two of its six electrons in the innermost orbit and the other four in the second orbit (which is the normal outer orbit for this atom). The copper atom with its 29 electrons is also included here. It's first three orbits are full and there is one lonely electron in its fourth orbit (which is the normal outer orbit for this atom).

Now ... here's where the quantum concept comes in. Sometimes an electron will jump to a lower orbit when nobody is looking. (Actually, electrons are obliged to do this whether anyone is looking or not, because of law 8 above.) Since a lower orbit means a lower level of energy, the electron (and hence the whole atom) gives up a specific amount (a quantum) of energy when it makes this jump — no more, no less. The actual amount depends on the chemical element (type of atom) and the two orbits involved in the jump.

Hang on, dude. Things can't just give up energy. The energy lost must go somewhere else, right? Like ... conservation of energy and all that ....

Well, that's exactly right, Tri-face. But I was coming to that. Honest.

The energy lost is emitted from the atom in the form of a photon — a quantum of electromagnetic radiation with exactly the same amount of energy given up by the hop-happy electron. And guess what ....

The same exchange of energy can work in reverse. If the atom gets zapped by a high flying photon that happens to have exactly the same amount of energy as the photon that was emitted in the first exchange (described above), this high flying photon gets absorbed by the atom (it disappears) and its energy goes into "exciting" the atom.

Exciting the atom? Well, I guess I'd get a little excited too if an energetic photon came along and zapped me one!

More specifically, the added energy the excited atom receives shows up as a jump by an electron from the same lower level to the same higher level as in that first exchange (described above).

As you will see in Chapter 9, these emissions and absorptions of photons at specific quanta of energy enable astronomers to identify the atoms (and even the specific electron orbits involved) when they analyze the light from stars and galaxies. And from that identification, astronomers can work out a lot of basic information about those stars and galaxies.

When you stop to think about it, this linking of the micro and macro worlds by astronomers is quite something:

By knowing how much energy is associated with a particular electron jump in the world of quantum mechanics, astronomers can determine what is going on in stars and galaxies millions of light-years away in the world of Newtonian mechanics.

This concludes our comparison of the macro and micro worlds.

Chapter 3 —
Distances from Here to There

To give you some idea of the size of the universe, Table 3-1 lists the distances of various objects from Earth, starting with the shortest and ending with the longest. The second column gives the distances in kilometers, but the numbers get so huge for really distant objects that they become kind of meaningless. So astronomers use a much larger unit of measurement that (a) keeps the distance numbers much smaller and (b) makes them more meaningful from a scientific point of view. This unit is much larger than the AU unit (Earth-sun distance) introduced in Chapter 1.

The Light-year

Astronomers express long distances in terms of **light-years** (abbreviated ly), which is the distance that light travels in one year. The third column in the table gives the distances in this unit.

One light-year represents a very large number of kilometers because light travels through space at almost 300,000 kilometers every second (if you can imagine anything going that fast). Nothing travels through space faster than light. Certainly not the family car. If you multiply that speed by the number of seconds in a year …

300,000 × 365.25 days × 24 hours × 60 minutes × 60 seconds

you get a whopping 9.46 trillion kilometers for one light-year.

As Table 3-1 shows, objects within our solar system are just a small fraction of a light-year away and therefore better expressed in kilometers or AUs or some fraction of a light-year, such as a light-second, a light-minute, or a light-hour.

 NOTE: It takes a scientific calculator to handle these large numbers. You can always access one on the internet or on many of today's handheld devices. You might, however, want to buy such a calculator at the store and keep it handy alongside this book. Although you certainly don't need one to read the book, you might find it fun to try some of the calculations yourself that pop up throughout these pages. (Besides, somebody should be checking my arithmetic!) Or, why not experiment and try some calculations of your own, using the reference data in the back of the book? Not only can you key in something like 10 to the power of 22, you can just as easily key in something as crazy as 37 to the power of 5.924. That's

$37^{5.924}$ or 37 multiplied by itself 5.924 times, which just happens to be … let's see now (tap tap tap) … 1,949,965,885.

And besides, you can impress all your friends with a store bought calculator like that and pick one up for a few dollars (the calculator that is, not the friend).

Looking into the Past

Because the light from any of these objects takes time to reach Earth, what you see is what the object looked like *at the time the light left it*. So when you look at the moon, you are seeing it as it was 1.3 seconds ago, not as it is now. When you look at the sun (with just a fleeting glance, otherwise it will damage your eyes), you are seeing it as it was 8.3 minutes ago. If the sun suddenly stopped shining, you wouldn't know about it for another 8.3 minutes.

Object	Distance from Earth	
	Kilometers (km)	Light-years (ly)
Moon	384 thousand km	0.000000041 (or 4.1×10^{-8}) ly (or 1.3 light-seconds)
Sun	150 million km	0.000016 (or 1.6×10^{-5}) ly (or 8.3 light-minutes)
Pluto	6 billion km	0.00066 (or 6.6×10^{-4}) ly (or 5.8 light-hours)
nearest star (Proxima Centauri)	40 trillion (or 4×10^{13}) km	4.2 ly
our galactic center	2.5×10^{17} km	26 thousand ly
Andromeda galaxy	2.4×10^{19} km	2.5 million ly
most remote galaxies seen	over 1.1×10^{23} km	over 12 billion ly

Table 3-1 Distances from Earth

In other words, when you look at celestial objects, *you are looking into the past.*

When you look at Proxima Centauri, the nearest star to us, the table tells us that you are looking 4.2 years into the past. Viewing the Andromeda galaxy is looking 2.5 million years into the past. The most remote galaxies that the Hubble Space Telescope has viewed takes us back in time more than 12 billion years when the universe was very young. Makes you wonder what those distant galaxies might look like today, but we have no way of knowing. They have moved much farther away during the intervening 12 billion years and their light does not reach us instantaneously.

Of course, this whole business of looking into the past applies to everything, not just celestial objects. When you look at someone across a living room, you are not seeing them as they are now. Good heavens, no. It takes time for the light reflecting off that person to reach you — about 0.00000003 seconds.

You might wonder how on Earth (literally) astronomers can determine how far away these distant objects lie. They don't have a measuring tape long enough to do an actual measurement. Even if they did, they don't have a spaceship to stretch the tape that far. And even if they had such a spaceship and it could magically travel at the speed of light, it would still take far too long to get there and back — five million years round trip for the Andromeda galaxy.

Just think how many lunches they would have to pack. Let's see now ... 5 million years x 365.25 days ...

As it turns out, determining these distances with some accuracy is one of the biggest challenges astronomers face. We will see how they do it when we get to Chapter 10.

Chapter 4 —
Empty Here, Crowded There

When you look at the sky on a clear moonless night in dark countryside, the universe appears quite full. Hundreds of individual stars appear in addition to that dense Milky Way band of stars mentioned in Chapter 1. Also, there are densely populated regions of stars here and there. One such region is in the central bulge of spiral galaxies. Another is in large spherical groupings of stars called globular clusters that exist on the outskirts of some galaxies, including our own.

A Universe That Is Almost Empty

Aside from these pockets of high stellar densities, however, most of the universe is pretty empty. Galaxies are separated by millions of light-years and that vast expanse of space between them is virtually empty.

My dictionary says **stellar** means "of or relating to the stars". Or "composed of stars". Our illustrious author uses it from time to time.

Distances Between Stars

Even within a galaxy, the stars are typically spaced far apart, although there is some gas and dust in the intervening space. In the last chapter we saw that the nearest star to our sun is hardly what you would consider a close neighbor. It's 4.2 light-years away! Here are a couple of scaled models (take your pick) that might help you visualize just how far we are from our nearest neighbor:

Model 1 — If the period at the end of this sentence represents our sun, then the nearest star would be represented by another period roughly the same size located (are you ready of this?) about 17 kilometers (ten miles) away. How's that for empty?

Model 2 — If a tennis ball represents our sun, then the nearest star would be represented by another tennis ball located about 1,920 kilometers (1,190 miles) away. That's roughly the distance from Boston to Miami. With no other tennis balls in-between.

Average Cosmic Emptiness

Overall, the universe is ridiculously empty. If we could magically redistribute all the existing matter evenly throughout the universe so that there were no galaxies, stars or planets — just a uniform density of matter everywhere — we would have what you see in Figure 4-1. We would have about three atoms for every cubic meter of space!!! Scientists can't make nearly that good a vacuum with their laboratory equipment. Since we know the approximate size of the atom, we can now calculate that the universe is roughly 99.99999999999999999999999 percent empty, based on Figure 4-1. That's 24 nine's after the decimal point.

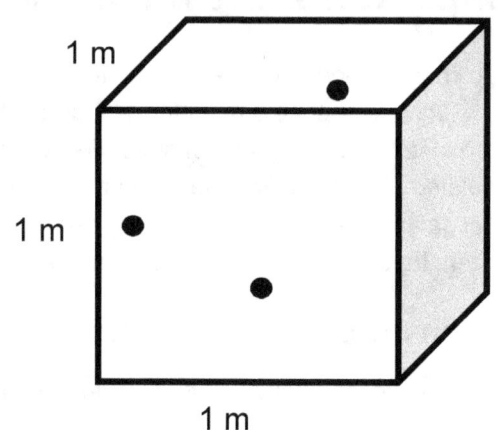

1 m

1 m

1 m

Figure 4-1 Atoms in Space

Atomic Emptiness

Finally, if we consider atomic structure, then the picture gets even more bazaar because the atom itself is not a solid structure. In fact, it is practically empty. As Figure 4-2 shows, the distance from the nucleus to the cloud of electrons whirling around it is roughly 10,000 times the diameter of the nucleus itself. As with most of the diagrams in this book, this one is not drawn to scale. But you knew that.

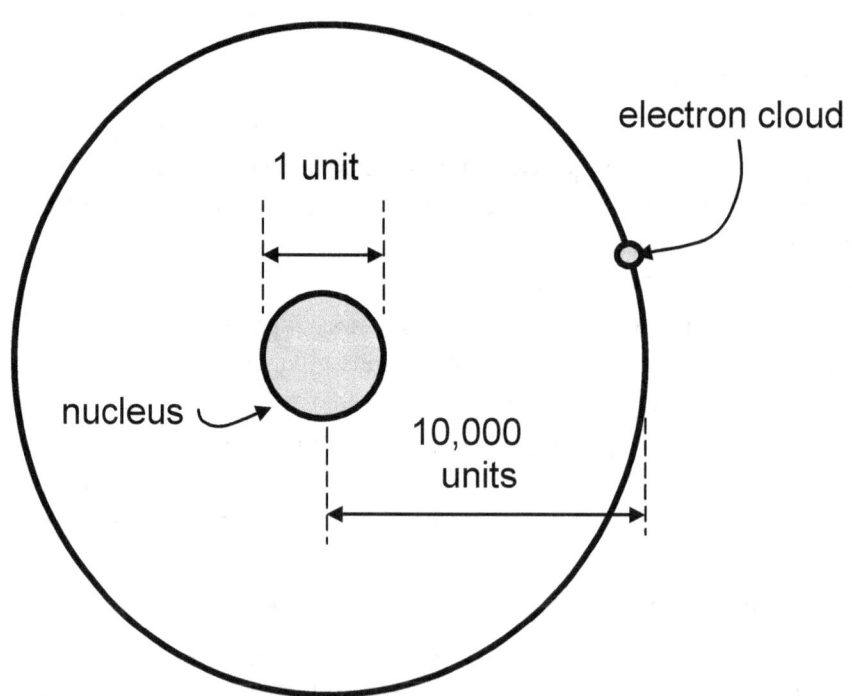

Figure 4-2 Space in Atoms

With these measurements, we can now calculate that the atom itself is roughly 99.999999999999 percent empty. That's 12 nine's after the decimal point.

The study of cosmology is easy — there's almost nothing to this stuff.

NOTE: If you want to try your hand at these sort of weird calculations with your shiny new scientific calculator (you did buy one, right?), you need to know that the volume of a sphere is equal to four thirds "pie" R cubed. "Pie" has nothing to do with food — it's the Greek letter pi and represents the number you get when you divide the circumference of any circle by its diameter. The formula for the volume looks like this:

$$V = \frac{4\pi R^3}{3}$$

where:

V = volume of the sphere

π = constant pi (3.14 rounded to two decimal places)

R = radius of the sphere.

So in Figure 4-2 the volume of the nucleus would be

$$V = \frac{4 \times 3.14 \times (0.5)^3}{3} = 0.52 \text{ cubic units}$$

It doesn't matter what the unit is because we're going to use the same unit to compare with the whole atom. The volume of the whole atom (including all that empty space between the nucleus and the electrons) would be:

$$V = \frac{4 \times 3.14 \times (10,000)^3}{3} = \text{a whopping } 4.2 \times 10^{12} \text{ cubic units}$$

Compare that with the 0.52 cubic units of matter in the center of the atom and you can see why the atom is roughly 99.999999999999 % empty.

Densities of Different Objects

Despite all this talk of emptiness, there are some things within the cosmos that are unimaginably dense. Table 4-1 shows some incredible variations in density when comparing different objects, starting with the lowest density and ending with the highest.

The density of an object — as everybody knows — is simply its mass divided by its volume:

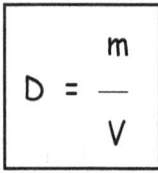

where:
D = density
m = mass
V = volume.

The densities reported in the table are in kilograms per cubic meter (abbreviated kg/m^3). Since everyone is familiar with water, the row in the table for fresh water is highlighted in gray for comparison with the other, less familiar, objects. If you have ever carried a full bucket of water any distance, you know that water is fairly heavy. Fresh water just above the freezing point has a density of 1,000 kg/m^3 (or 64 lb/ft^3 for those who prefer that antiquated system).

Although water is very much denser than air, you can see that it is less than one fifth the average density of planet Earth. We think of gas as being very light and, of course, the sun is gaseous (being too hot for any solids or liquids to exist there). Yet the sun is denser than water because it is so massive that its internal gravity pulls all that gaseous matter tightly together.

You may find the last two densities listed (from the remains of dead stars) beyond belief. How on Earth (literally) could a coffee mug of this stuff possibly weigh hundreds or billions of tons? We'll find out in Chapter 14, but the clue is in Figure 4-2. Stay tuned.

OK, OK ... so maybe I was a little hasty with my previous comment about this stuff being easy.

Objects	Density (kg/m^3)	Comment
interstellar space	10^{-25} to 10^{-15}	contains various amounts of gas and dust
Earth's atmosphere	1.2	at 20° C at sea level
fresh water	1,000 (that's 64 lb/ft^3)	at 0° C (that's 32° F)
planet Earth	5,515	average density of the whole planet
lead	11,340	atomic number 82
Earth's inner core	13,000	solid iron under pressure of Earth's mass above it
the sun	1,410	average density of the whole sun
the sun's core	160,000	gas (mostly H and He) under pressure of the sun's mass above it
white dwarf star	10^9 (1 billion)	shrunken remains of a low-mass star like the sun after it has burnt out and gravitationally collapsed (one coffee mug of it would weigh* 500 tons on Earth)
neutron star	4×10^{17}	shrunken remains of a high-mass star after it has burnt out and gravitationally collapsed (one coffee mug of it would weigh* 50 billion tons on Earth)

* Mass and weight are two different things. Mass is the amount of material, whereas weight is the gravitational pull on that mass. So the weight of an object depends on the strength of the gravitational field in which the object is located. For example, you would weigh about six times less on the moon than you do here on Earth.

Table 4-1 Density of Things

Chapter 5 —
Cold Here, Hot There

We have already seen some astonishing cosmic differences when looking at distances and densities. Temperature is just one more thing to add to the list. Table 5-1 shows some incredible variations in temperature when comparing different objects, starting with the lowest temperature and ending with the highest.

The temperature of any substance reflects the speed at which the atoms or molecules are jiggling around within it. The faster they jiggle, the higher the temperature. The coldest possible temperature is called **absolute zero** where the atoms are barely moving. Being very restless, atoms can never completely stop jiggling.

The temperatures in the table are shown in three different temperature scales, just to keep everybody happy:

Kelvin — because that's what astronomers like to use. On this scale, zero represents absolute zero, so there are no negative Kelvin temperatures.

Celsius — because that's what most of the world uses these days. On this scale, absolute zero is minus 273 degrees. Celsius has a metric flavor with 0 degrees for the freezing point of water and 100 degrees for the boiling point of water.

Fahrenheit — because it is still used in the USA. On this scale, absolute zero is minus 460 degrees with 32 degrees for the freezing point of water and 212 degrees for the boiling point of water.

NOTE: To change a temperature from Celsius to Kelvin, you simply add 273 to the Celsius figure. So water freezes at 273 K and boils at 373 K. At the very high temperatures found inside stars, this 273 degree difference in the two scales doesn't matter much.

Table 5-1 takes us from almost absolute zero in deep space all the way up to over ten billion degrees in the core of massive stars just before they die in a supernova explosion (see Chapter 14). The massive star presented in this table is 25 times more massive than our sun. It is referred to as a "25 solar mass star" and denoted by astronomers as a 25-M⊙ star, where the symbol ⊙ represents our sun.

Below the table, Tri-face is trying to tell us how spoiled we are, living in Earth's protected environment.

On a final note to this short chapter — there is something odd in the table. Even though Mercury is closer to the sun than Venus, it is not as hot as Venus. How come? The answer can be found in the Comments column. No wonder scientists worry about potential increases to green house gases here on Earth.

| Cosmic Objects | Average or Approximate Temperature (unless otherwise stated) | | | Comments |
	Kelvin (K)	Celsius (°C)	Fahrenheit (°F)	
outer space	3	-270	-454	only 3 degrees K above absolute zero
surface of Pluto	50	-223	-369	very approximate
surface of Mars	Min: 133 Max: 293	Min: -140 Max: 20	Min: -220 Max: 68	
surface of Earth	Min: 183 Ave: 293 Max: 333	Min: -90 Ave: 20 Max: 60	Min: -130 Ave: 68 Max: 140	
surface of Mercury	Day: 623 Night: 103	Day: 350 Night: -170	Day: 662 Night: -274	the planet's days and nights are very long
surface of Venus	750	480	900	runaway green house gas effect (lots of CO_2)
surface of the sun	5,800	5,527	9,981	
core of the sun	15.5 million	15.5 million	27.9 million	hot enough for nuclear reactions (see Chapter 13)
core of 25-M⊙ star	40 million	40 million	72 million	more than hot enough for nuclear reactions (see Chapter 13)
core of dying 25-M⊙ star	10 billion	10 billion	18 billion	just before exploding as a supernova (see Chapter 14)

Table 5-1 Temperature of Things

Meanwhile ... we folks, who live on a planet with an average surface temperature of 20°C, complain if the temperature drops below 10° C or rises above 30° C.

Chapter 6 —
Older Than Your Great Great Grandmother

The previous chapters gave us a glimpse of what the universe looks like. But has it always been like this or did it have a beginning? And if so, when? Like, how old is it? And will it have an ending? And if so, when? Let's start with our own cosmic neighborhood and work out from there.

Our Solar System

As we shall see in Chapter 9, astronomers can determine an awful lot about the stars by analyzing the light that comes from them — chemical composition, luminosity, surface temperature, and such things. From this they can estimate the age of individual stars. It turns out that our sun is about 4.5 billion years old. This figure is in good agreement with the age of meteorites found on Earth, as determined by radioactive dating techniques. Since meteorites represent original material from the formation of the solar system, the 4.5 billion year figure is considered an excellent estimate for the age of the solar system.

Our Milky Way Galaxy

The age of our galaxy has not been determined with the same certainty. Recent studies of some very old stars in the halo that surrounds the galaxy suggest an age of over 13 billion years. The thin disk of the galaxy is considered to be more like 10 billion years old. So it may be that the galaxy's formation started when the universe was still very young.

The Universe

In the early part of the last century, most scientists and philosophers thought that the universe had always been here — static and unchanging — that it was just one big continuous expanse of stars that formed what we now know to be the Milky Way galaxy. Even Albert Einstein included this thinking in his theory of general relativity in 1915.

But the work of American astronomer Edwin Hubble in 1928 changed all that when he turned the 100 inch Mount Wilson telescope toward various nebulae in the night sky. Back then, the word "nebula" meant any celestial object that appeared fuzzy and extended in a telescope. In other words, not just a single point of light like a star. Some of these objects turned out to be clouds of gas and dust and others were clusters of stars — all within our galaxy. But others turned out to be huge islands of stars at distances way beyond our galaxy. This was, indeed, a mind-blowing discovery!

One really large nebula turned out to be the Milky Way's near neighbor — the Andromeda galaxy (see Figure 1-3 on page 8). Other nebulae were at much greater distances. Hubble's observations can be summed up as follows:

- There were "island universes" (galaxies) far beyond the Milky Way.
- With the exception of a few nearby galaxies like Andromeda, these other galaxies were moving away from us.
- The farther away they were, the faster they were receding.
- The matter in the universe was expanding at a measurable rate.

Hubble came to these conclusions by determining the approximate distance of each galaxy and the velocity at which it was receding from us. In Part II we will show you how astronomers go about determining such things.

Hubble's observations showed that the receding velocity of far away galaxies was directly proportional to their distance. So a galaxy at twice the distance of a closer galaxy would be receding at twice the velocity and one at three times the distance would be receding at three times the velocity.

In mathematics this is called a **linear** (or straight line) **relationship**. If you plot distance (d) against recessional velocity (v) for each far away galaxy on a graph, you get something like Figure 6-1 where the plotted points suggest a straight line relationship between the values represented on the X and Y axis.

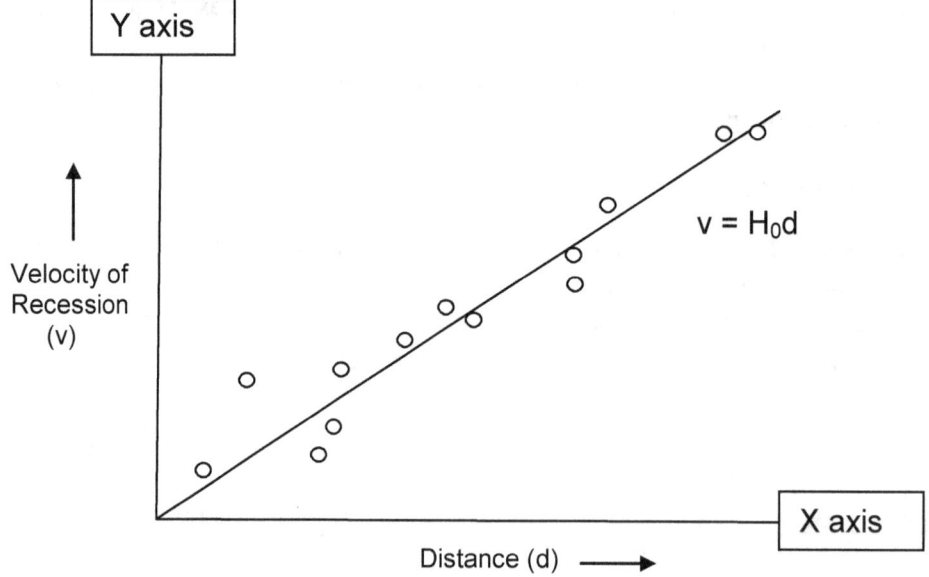

Y axis

Velocity of
Recession
(v)

$v = H_0 d$

Distance (d) ⟶

X axis

Figure 6-1 Hubble Law

The only reason the plotted points don't fit perfectly on the straight line is because their values have been determined by observation and are therefore not 100% accurate. Astronomers have a pretty accurate way of determining recessional velocities, but not so for distances. The degree of accuracy is much less for distances.

The relationship shown in Figure 6-1 is known as the Hubble law and is expressed by the following simple equation:

$$v = H_0 d$$

where:
v = recessional velocity of a galaxy
H_0 = Hubble constant
d = distance to the galaxy.

At this point, don't worry about what units to use for velocity and distance. We will have a closer look at this law in Chapter 11, where we will see just how fundamental it has been for those cosmology geeks who study the expansion of the universe.

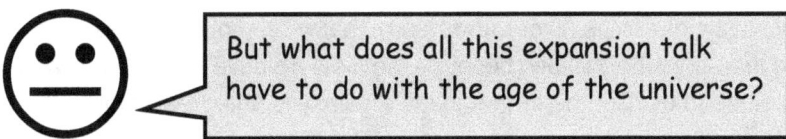

But what does all this expansion talk have to do with the age of the universe?

I was just coming to that, Tri-face.

If all the remote galaxies are moving away from us, they must have been a lot closer in the past. If we could run the cosmic clock backwards far enough, it seems reasonable that all the matter in the universe would come together at some point in the distant past. If that's the case, something really spectacular — really awesome — must have happened back then to send all this matter and energy hurtling outward with such enormous force. Like some kind of (you guessed it) **big bang**.

This whole scenario suggests there must be a way to calculate just how much time would be needed to get back to that "big bang" event. That duration would then represent the age of the universe. In Chapter 11 we will see how the Hubble law enables us to do this calculation.

But not to keep you in suspense about the answer — cosmologists have recently determined with considerable confidence that the universe is 13.7 billion years old. But you probably knew that.

NOTE: This answer is actually based on a more accurate technique than the Hubble law. It's based on the analyze of the cosmic microwave background radiation that exists throughout space today and that originated when atoms first formed in the infant universe (as discussed in Chapter 12).

> HANG ON A MINUTE! I have a problem here. If the universe is expanding, how come a galaxy like Andromeda is moving toward us, not away from us?

Yeah, good point Tri-face. I didn't want to get into this, but you've forced my hand. The expansion we are talking about here is actually the expansion of *space itself*, not the movement of galaxies *through* space. All the galaxies (including our own and Andromeda) are being carried along with this expansion because they are part of space itself. Sometimes this expansion is referred to as the **Hubble flow**. Think of these galaxies as canoes without a paddle on a fast flowing river. In a sense, the canoes become part of the river as they get carried along with the current.

Having said that, however, galaxies also move *through* space with their own local motion. This is because they have a mutual gravitational attraction to other galaxies. For galaxies close to us (like Andromeda), that local motion can be much more dominating than the Hubble flow, because that flow (relative to us) is not very great for objects close to us. But the farther away a galaxy is, the greater the velocity of this Hubble flow (relative to us) — as the Hubble law shows — so it eventually dominates over this local motion.

In other words, because of the local motion, the Hubble law doesn't apply very well to nearby galaxies like Andromeda. But that doesn't make it any less important as a cosmic discovery.

> OK, but if that's the case, then we humans must be at the center of the universe because these remote galaxies are all moving away from us no matter what direction we look in. Right, dude?

Well actually, no. We're not at the center. This is another difficult point I was hoping you wouldn't bring up. When astronomers are asked about this, they tend to return to their childhood and start talking about balloons with dots painted on them. Imagine blowing up such a balloon, as shown in Figure 6-2.

Imagine that the surface of this inflating balloon is our expanding universe. (Just the *surface*, not the inside of the balloon.) And imagine that each dot represents a galaxy. As the balloon gets bigger the dots move farther and farther away from each other. If you were an ant sitting on one of those dots (say one representing our Milky Way galaxy), it is true that all the other dots appear to be moving away from you no matter in which direction you look. And the farther a dot is from you on this surface, the faster it seems to be moving away. You certainly appear to be in the center of all that expansion.

Figure 6-2 Expansion of Cosmic Space

But what if another ant is sitting on a different dot? What will it observe? You got it — exactly the same thing. That ant too will see itself as being at the center of the expansion. The same goes for any dot you care to pick. Every ant on every dot has an equally valid argument for claiming the central position.

Well, obviously they can't all be at the center or we'd have more than one center. Therefore, none of them are at the center. When you think about it, the *surface* of the balloon (which represents the expanding universe in this analogy) has no center. And so, cosmologists will tell you that the universe doesn't work the way we naturally think it should. It doesn't have a center.

These scientists tell us that the expansion of the universe is like the balloon surface analogy, but in three dimensions of space instead of just the two that the balloon surface represents. Now that's just about impossible for us ordinary humans to visualize, but these cosmology types can describe it mathematically.

The next time you find yourself hanging out with a bunch of cosmologists you might want to quiz them on this particular point. It's a bit beyond me.

Chapter 7 —
And Always Changing …

A s you can see from the last chapter, things do change over very long periods of time. But we humans, with our short lifespan, tend not to think about such changes. We talk about the "fixed stars" because they haven't changed their positions in the sky for centuries. Or so it appears. We tend to forget that a century is but a blink of an eye (as they say) in the life of the universe.

Earth is 4.5 billion years old (about a third as old as the universe itself), but the human race has only been kicking around for the last million years or so (depending on how you define the human race). A million years is a mere 0.02 percent of our planet's age. And an 80 year old person has only been around for 0.008 percent of that last million.

> I wish you wouldn't dwell on things like that, dude. It makes me realize just how utterly insignificant I am in the scheme of things — along with everybody else, of course.

Sorry about that. Didn't mean to get you all depressed and philosophical.

So we have to look to the cosmologists and their evolutionary picture of the universe to get some idea about cosmic change. We have to think in terms of millions and billions of years. This chapter briefly describes some of these changes, starting close to home, then moving farther out.

Planet Earth Does Not Go Topless

Back in Chapter 1 we mentioned that most celestial objects spin like a top. And our planet is no exception, completing one rotation in slightly less than 24 hours. What we didn't mention is that— like any top — its axis tends to wobble during the spin. This wobble causes the north and south ends of the axis to move in a small circle that makes one complete revolution every 26,000 years. Astronomers call this motion **precession** — just one of the "lesser motions" we didn't talk about in Chapter 1.

Today, the north end of Earth's axis points directly to the star Polaris, which is why Polaris is known as the North Star. But this situation will gradually change as the axis slowly points elsewhere on it 26,000 year circular journey. Some 12,000 years from now (around A.D. 14,000) the so called "North Star" will be the bright star Vega.

The Earth-Moon Dance

It seems nothing is perfectly constant, not even within the Earth-moon system. The complex play of gravity between the moon and Earth that creates the tides is causing the moon's orbit to gradually get bigger and Earth's daily rotation to gradually slow down. The appropriate word here is "gradually". The moon is moving farther away from Earth by a mere 3.8 centimeters every year. That's 1.5 inches. And Earth's day is getting longer by two milliseconds every century. That's just two thousands of a second.

These changes may not be enough to adjust your moon calendar or your watch, but over millions of years they do have an effect. For example, take the glorious sight of a total eclipse of the sun that we enjoy today — when the moon's disk fits perfectly over the sun's disk and displays that classical flare of the sun's corona around the dark lunar disk (see Figure 7-1). This is only possible because both disks are practically the same size as viewed from Earth. Many millions of years ago it didn't look like that. Nor will it look like that many millions of years from now.

Figure 7-1 Total Solar Eclipse

Why not? Because in the distant past the moon was much closer to Earth and therefore much bigger in the sky than the sun. It blocked the sun's light for much longer than today's eclipses and didn't permit the spectacular display of the sun's corona behind it. In the distant future, the moon will be much farther away and therefore much smaller in the sky than the sun. It won't be able to cover all of the sun's disk — so no more total solar eclipses.

It seems we humans live in a fortunate time and place, cosmologically speaking. What are the odds of an intelligent civilization emerging on a planet just at a time when that planet has a moon with the same apparent diameter in the sky as its sun? How cool is that?

Mass Extinctions (poof go the dinosaurs)

As a civilization, we constantly deal with natural disasters such as earth quakes, floods, droughts, hurricanes, and so on. We don't, however, think in terms of mass extinctions. You may have read somewhere that life on Earth has experienced a few such extinctions in the history of the planet when many (if not most) species died out, never to return. The most likely cause for such extinctions is the arrival of a large asteroid or comet (ten or more kilometers in diameter), hurtling through Earth's atmosphere at great speed. The energy produced by such an impact generates a deadly force measured in the thousands of megatons of TNT, bringing widespread destruction around the globe.

Just ask the dinosaurs. They were wiped out 65 million years ago when a huge asteroid plunged into the Gulf of Mexico. This sort of thing could happen again, although it's more likely our own mismanagement of the planet will wipe us out first. Sixty five million years is a long time compared to the history of the human race.

The Life Cycle of a Star (poof goes the sun)

As we shall see in Part II, stars are not permanent fixtures in the sky. They are born out of the surrounding gas and dust and eventually die — lights out, so to speak. And so it will be with our own sun. As described in Chapter 13, our sun is now middle aged, but will continue to get hotter as it has done throughout its life. In five billion years or so, it will run out of its nuclear fuel and take Earth with it in a fiery death.

Now, that's what I call real change!

The "Fixed Stars" Aren't Fixed

In addition to their merry-go-round trip around the galaxy along with all their neighboring stars, individual stars do have an independent motion of their own. They move relative to neighboring stars. Astronomers call this motion **peculiar motion** (but then, as noted earlier, astronomers are a peculiar bunch). Over several thousand years the appearance of some constellations may change slightly. Certainly nothing to worry about.

Fasten Your Seat Belt — Here Comes Andromeda

As you might expect from Newton's universal law of gravitation, galactic collisions are common throughout the cosmos. Astronomers have observed ample evidence of this. At the current rate that Andromeda and our own galaxy are moving toward each other (120 km/s), we may be involved in a galactic collision a few billion years from now. Using this velocity and a distance of 2.4×10^{19} km between the two galaxies (see Table 3-1 on page 26), we can calculate the amount of time before this potential collision occurs:

$$\text{time} = \frac{\text{distance}}{\text{velocity}} = \frac{2.4 \times 10^{19} \text{ km}}{120 \text{ km/s}} = 2 \times 10^{17} \text{ s}$$

That's a lot of seconds, so you might want to divide them by the number of seconds in a year to get the answer in years:

$$\frac{2 \times 10^{17} \text{ s}}{365.25 \text{ days} \times 24 \text{ h} \times 60 \text{ min} \times 60 \text{ sec}} = 6.3 \times 10^9 \text{ years}$$

That's 6.3 billion years, but the time would no doubt be less than that. Why? Because the gravitational attraction between these galaxies will increase as the distance between them decreases in accordance with Newton's universal law of gravitation back on page 10.

If we're still around at the time, we would have a front row seat to watch the action. Nobody knows for sure if this will happen, because this measured rate is only the **radial** velocity of the galaxy's motion — that component of velocity that is moving *directly* toward or away from us.

The other component of galactic motion is the **transverse** velocity — that component that is moving left (or right) and up (or down) from the direct line between the galaxy and us. Think of this as its sideways velocity relative to us. Unlike the radial component of velocity, astronomers cannot measure the sideways component directly. They can only make a rough estimate by indirect methods. Because of this, the likelihood of a collision is not clear. Place your bets.

The good news is that such a collision would not be as catastrophic as it sounds. You have already seen that the distances between stars are very great, making galaxies — in a sense — almost empty. So the chance of a collision between a star in one galaxy and a star in the other galaxy is extremely small. The two galaxies would simply merge, creating a completely different looking night sky. It would be a spectacular show for any civilizations that may be hanging around from either galaxy at that time. Mind you, this would not exactly be a fast action movie, since it would take place over millions of years.

Expansion of the Universe (poof go the galaxies)

As the Hubble law indicates, the velocity of the expansion of space increases with distance. The farthest regions of the universe (beyond our observational limit) are expanding faster than the speed of light. Eventually, all the galaxies in our local ...

HEY, NOT SO FAST! With all due respect, dude, you told your readers back in Chapter 3 that nothing could go faster than light. Remember?

Ah yes, but the statement made there was, "Nothing travels through space faster than light." The key words here are *through space* — the kind of motion we all know and understand. This statement puts no restriction on how fast space itself can move with the expansion of the universe. Cosmologists tell us that there is no speed limit on this expansion.

Now ... what were we talking about before Tri-face interrupted? Oh yeah ... eventually, all the galaxies in our local group might merge together into one giant galaxy (who knows), but the galaxies well beyond our local group would continue to recede farther and farther.

Here's another one of those Carl Sagan type number-of-stars-in-the-universe statements for you to ponder. In his book *NightWatch*, Canadian amateur astronomer and writer, Terence Dickinson, claims that *the universe will expand by another 100 trillion cubic light-years in the time it takes to read that sentence in his book.* Impossible to grasp? Yep, pretty much. It's a big universe and getting bigger by the minute.

Final Destination of the Universe

If this expansion continues forever, the galaxies will become so spread out that future civilizations won't be able to see them, no matter how powerful their telescopes. The cosmos will become very cold, dark, and even emptier than it is now.

As Woody Allen once said, "eternity is very long, especially towards the end."

However, there is hope. The gravity associated with the matter in the universe is working against the force of the cosmic expansion, trying to pull everything back together again. It's a tug of war. If there is enough cosmic matter to eventually bring the expansion to a halt, the universe could then start contracting. In that case, it would eventually collapse to the same unimaginably dense state that it started with at the time of the Big Bang. Not that we humans would be around to see it.

So the universe might reverse it's history and end in a Big Crunch. What then? Another Big Bang? Until recently, cosmologists thought that the ultimate fate of the universe depended mainly on how much matter it contained. As we shall see in Part III, that simple picture has become more complicated. Stay tuned.

Well, here's a depressing thought. If gravity does not come to the rescue, neighboring galaxies will drift too far apart to be seen billions of years from now. Any intelligent civilization living at that time, *would have no way of knowing what we know today!!* The evidence would be gone. Those folks would not know that they are living in an enormous expanding universe with billions of other really remote(!) galaxies and that it all began with a Big Bang hundreds of billions of years ago.

Well, Tri-face, I gotta admit that's an awesome thought you have there. Very profound observation. I'm quite impressed.

Part II

The Science of
Astronomy and Cosmology

Chapter 8 —
Theoretically Speaking ...

In Part I we presented some basic facts and figures about the cosmos. We now want to go one step further and introduce you to the science that lies behind these facts and figures. It's a fairly big step, but if you truly enjoy the intellectual challenge that science offers, you should find this new adventure a lot more fun than just reading facts and figures.

In Chapters 9 to 11, we will give examples of how astronomers and cosmologists actually use science in their work. We will explain the relevant science and math in small easy steps and then show how it is used to determine specific information about the cosmos. Don't worry about the math. We're not talking logarithms or calculus here. Just a few really simple algebraic equations. As we pointed out in Chapter 1, you can't learn much science without getting into some math. I guess that's good news or bad news, depending on whether or not you like math.

In Chapters 12 to 14, we will lean more heavily on theory to tell you about the evolution of the universe and what goes on inside stars. Theories come under different labels like "theory", "law", or "model" — but it's still theory. For example, in Part II we mention the Hubble *law*, the Big Bang *theory*, and the Standard Solar *model*.

At this point, it's important to understand what scientists mean by the word "theory". Non-scientists tend to use the word very loosely. When they say, "Yeah, but it's just a theory", they tend to mean that it's just a lot of guesswork with nothing to back it up.

Hmm. That sounds like something I would say.

Well, smarten up, Tri-face. You should know better.

The Scientific Method

To explain "theory", we need to describe the **scientific method** — the basic procedure that scientists use to investigate things. They make observations. They record data. From this information, they come up with a set of ideas called an **hypothesis** that attempts to explain whatever it is they're studying. They then subject this hypothesis to experimental tests and further observations. If it survives this additional investigation without being disproved, then at some point it is considered to be a **theory**. The theory is then used to make predictions, which can then be tested in order to either (1) provide further evidence that supports the theory or (2) provide new evidence that shows the theory is flawed.

Of course, this procedure does not guarantee that any established theory is true. The key point here is that you can't prove that a theory is true. Something might come up later to demonstrate that under certain conditions the theory fails. The scientific method can only prove through investigation and testing that a theory is false. If found to be false, it is either modified or discarded and replaced by a better theory. This is how the advancement of scientific knowledge works.

Examples of Theories, Old and New

Discarded theories from the distant past include things like:

- a flat Earth
- a solar system with Earth at the center
- planetary motion based on perfect circles.

We humans used to think we were at the center of the universe. We now know that we are not at the center of anything. In the remaining pages of this book you will see examples of more recent theories having to be modified because of the latest findings.

Now ... Tri-face has asked if he can tell the story about a little old lady's version of the flat Earth theory. It's a story presented at the beginning of Stephen Hawking's 1988 book, *A Brief History of Time.* Heaven only knows if there is any truth to this story. But what the heck, I figure we could all use a laugh about now.

Thanks, Dude. Here goes: A well-known scientist (some say it was Bertrand Russell) once gave a public lecture on astronomy. He described how the earth orbits around the sun and how the sun, in turn, orbits around the center of a vast collection of stars called our galaxy. At the end of the lecture, a little old lady at the back of the room got up and said: "What you have told us is rubbish. The world is really a flat plate supported on the back of a giant tortoise." The scientist gave a superior smile before replying, "What is the tortoise standing on?" "You're very clever, young man, very clever," said the old lady. "But it's turtles all the way down!"

Sometimes a theory is proven wrong because it fails to satisfy tests or observations that involve extreme conditions. In such cases, the theory may still be considered valid within "normal" conditions and so it continues to be used within that limitation. A striking example of this is Newton's universal law of gravitation. It's now time to tell you about the dramatic downfall of this great theory. In so doing, it will force us to introduce the subject of relativity. Any book on cosmology would be cheating the reader big time if it didn't mention the role of relativity. And Tri-face and I certainly wouldn't want to cheat anybody.

Newton's Gravity Versus Relativity

When we introduced Isaac Newton's universal law of gravitation way back on page 10, we left out one small detail. This law is fundamentally wrong! It's out of date. Sorry about that.

Einstein's general theory of relativity, published in 1915, provides a more realistic picture of gravity and has been supported by certain astronomical observations. So why is Newton's law still used today? Because (a) it's a heck of a lot simpler to work with than Einstein's complex field equations and (b) it provides excellent accuracy for most physical conditions.

Cosmologists, on the other hand, deal with conditions totally unlike our everyday experiences —conditions of unbelievably strong gravitational fields and velocities approaching the speed of light. They have to use Einstein's relativity to get the right answers. Newton's law fails under these extreme conditions.

When the German physicist, Albert Einstein, started thinking critically about such things as space, time, mass, and light, he came to a startling conclusion. He found that space and time were so interconnected that he had to treat them as one single physical entity (or thing) — referred to as **spacetime** — rather than two separate entities — space *and* time. An event in spacetime can be specified mathematically by four coordinates (or numbers) to nail down its location: three to give its position in space and one to give its position in time.

Einstein no longer pictured matter as something that created a mysterious force acting instantaneously at a distance (Newton's gravity). He pictured matter as something that created a curvature in spacetime itself. The more massive the object, the greater the curvature. Of course, it's pretty hard for us mortals to visualize four dimensional spacetime, especially when it's curved. However, it can be described and dealt with mathematically using Einstein's field equations (although not by this author!).

As a simple analogy for spacetime, picture it as a two-dimensional rubber sheet stretched between some supports. Now place a heavy object on it like a bowling ball or a curling stone. You can imagine how the resulting weight would stretch the sheet (spacetime) downward to form a curved depression, as shown in Figure 8-1.

Scientists call that depression in spacetime a **gravity well**, because its curvature causes moving objects to spiral down into it. Imagine placing a marble on the rubber sheet near the bowling ball. The marble would move down the depression toward the ball. That's what objects do in real spacetime.

Newton would claim that the marble was being attracted to the ball by a force (gravity) between them that somehow acted instantaneously at a distance. Einstein would claim that the marble was simply being forced to follow

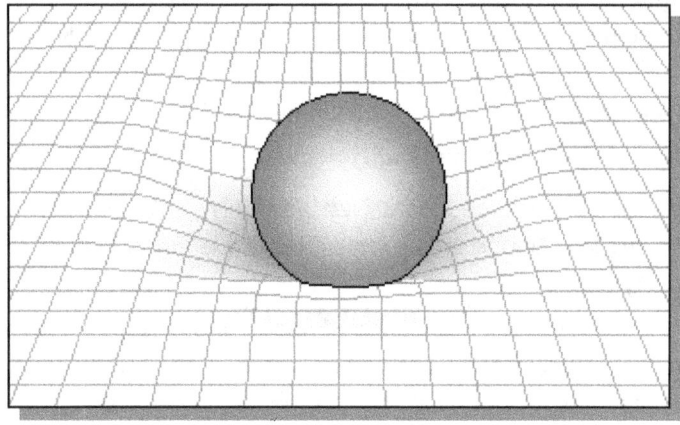

Figure 8-1 Matter Curving Spacetime

the curvature of spacetime. Every celestial object, be it a moon, planet, star, or galaxy, creates a curvature of four-dimensional spacetime around itself.

It is said that *matter tells spacetime how to curve and spacetime tells matter how to move*. But spacetime also tells *light* how to move, because light also has to follow the curvature of spacetime. That's why light is affected by Einstein's gravity even though the carriers of light (the photons) have no mass. Sir Isaac would have been dumbfounded to learn this.

This changes our view of the universe considerably. Instead of the more simplistic macro world versus micro world, we could break down the universe somewhat differently:

- the world of the very large described by relativity
- the world of the very small described by quantum mechanics
- the in-between world described by Newtonian mechanics.

Figure 8.2 is a picture of Einstein, the genius behind relativity. With all of his heavy (ha ha) thinking about gravity, it's no wonder the poor guy never had time for a haircut.

Figure 8-2 Albert Einstein

Well, that may be, Tri-face, but Albert did have a human side. And certainly a sense of humor. Among the many quotes attributed to this great man, one of my favorites involves humanity:

"Only two things are infinite, the universe and human stupidity, and I'm not sure about the former."

Chapter 9 —
The Spectrum of Light

We're hoping that Part II of this book will give you enough insight about the science of astronomy and cosmology to save you from making the same mistake that Auguste Comte made in the early 1800s. He was a French philosopher who argued that humanity would never know the nature and composition of the stars because they were so far beyond our Earthly reach.

 Boy, was he wrong! He obviously had no idea what a bit of starlight could tell us about its source.

Two hundred years after Comte, we now know the basic properties of light and matter and, yes, we do know "the nature and composition of the stars." We also know that stars and galaxies are made of the same stuff found right here on Earth. And finally, we also see that the physical laws we have uncovered here on Earth apply throughout the universe.

But I'm getting way ahead of myself. What is light, anyway? It's important to spend some time here dealing with this question because the physical properties of light are so fundamentally important to astronomy.

 **Advisory Bulletin about Chapters 9 to 11**: Our illustrious author gets a little technical in these three chapters. If you can take it slow, hang in tough, and dig these chapters, you should find them pretty interesting. And the rest of the book will be a snap for you.

Let There be Light

Light is energy in the form of radiation. A "light beam" is a stream of tiny particles — little packets of energy called photons that travel in a wavelike motion. So light behaves both like particles *and* waves. It took a few decades for a lot of big-name scientists to accept the reality of this strange, dual nature of light.

The Speed of Light

Photons of light have no mass but they zoom along at … well … the speed of light. According to modern measurements, that's 299,792.458 kilometers per second in a vacuum. In transparent mediums such as air, water, or glass, this speed is somewhat less. For most purposes, astronomers use a value of 300,000 km/s in their calculations because:

- this is extremely close to the actual value
- interstellar space is extremely close to a vacuum
- the other figures used in their calculations won't be that accurate anyway
- and finally, it's a heck of a lot easier to key in 300,000 than 299,792.458.

For a long time nobody (including Galileo and Newton) was able to measure the speed of light by means of earthly experiments. Everyone assumed it must be almost instantaneous. The first rough estimate was made by a Danish astronomer, Olaus Romer, in 1676 after he had been observing orbits of the moons of Jupiter over a period of several years. (He probably had a lousy social life.) These were the very same moons that Galileo had discovered in 1610 with an incredible invention called the optical telescope.

The timing of their eclipses with Jupiter was not what Romer expected. Oddly enough, these timings seemed to depend on the relative positions of Earth and Jupiter at the time of the measurements. In a eureka moment, Romar realized that these puzzling results could be explained if the light from his observations needed significant time to travel from Jupiter to Earth. Figure 9-1 illustrates this situation.

I don't know about you readers, but I'll probably never have a eureka moment. It must be pretty cool.

When Earth was farthest from Jupiter during Earth's annual orbit around the sun, the timing of his observations was about 16½ minutes later than when Earth was nearest to Jupiter at the opposite time of the year. So the extra 16½ minutes was needed for the light to travel the diameter of Earth's orbit, a distance not well known in Romar's time (hence his "rough estimate" of the speed of light). If you care to make the calculation, using the average Earth-to-sun distance of 150 million kilometers, and Romar's approximate time difference of 16½ minutes, you will get roughly 300,000 km/s for the speed of light. Give it a try. Speed is simply distance divided by the time to travel that distance. But remember not to confuse radius (1 AU) with diameter (2 AU) when working with Earth's orbit. You need the diameter for this calculation.

NOTE: I really shouldn't be talking "radius" and "diameter" when discussing Earth's orbit, since it is not a circle. As mentioned in Chapter 1, it's an ellipse. But it's so close to being a circle, I figure it's easier for all of us if I just pretend it's a circle, instead of mathematically defining an ellipse and all its components.

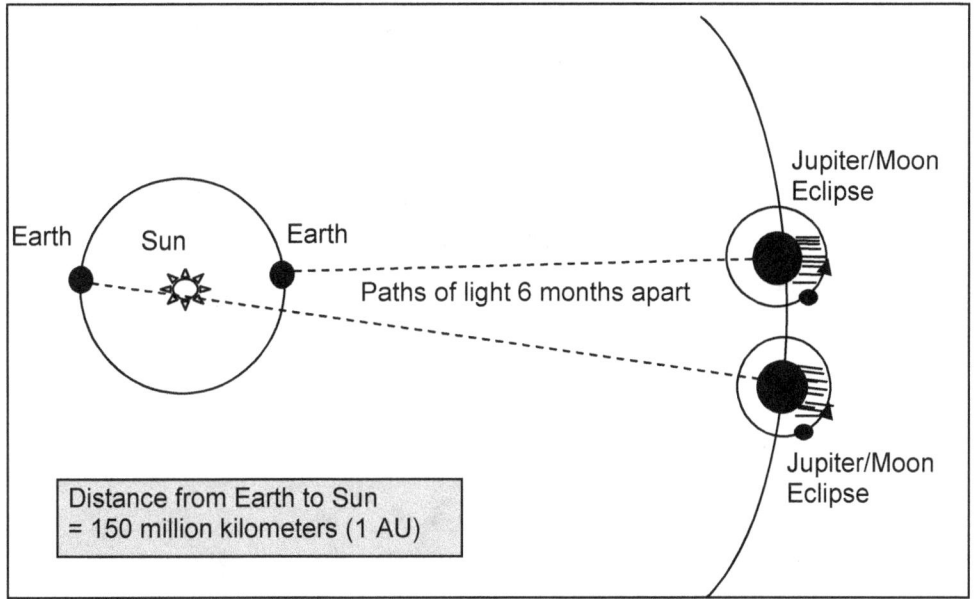

Figure 9-1 Romer's Observations of Jupiter/Moon Eclipses

The Wavelike Nature of Light

As you can see from Figure 9-2, a wave has an up and down oscillating motion as it travels forward in some specific direction (shown by a time arrow in the diagram). The two things about waves that are of most interest to astronomers are **wavelength** and **frequency**.

Wavelength is the length between one wave crest and the next and is symbolized by the Greek letter λ (lambda) in most textbooks. But we'll use the good old English letter w (for <u>w</u>avelength). Because these lengths are very short in the case of light, they are normally measured in some tiny fraction of a meter such as the nanometer (abbreviated nm) which is a billionth of a meter.

An oscillation is one complete up and down cycle of a wave motion and the frequency of that motion is the number of oscillations (or cycles) per second as the light passes a given point. Frequency is symbolized by the Greek letter v (nu) in most textbooks. But we'll use the good old English letter f (for <u>f</u>requency). A frequency of one cycle per second is known as a hertz (abbreviated Hz). But because frequencies are typically very high, they are usually measured in megahertz (abbreviated MHz) which is a million hertz.

Hey dude, a million hertz sounds awfully painful (ha ha). But why are you turning your back on those classical Greek letters? The scientific community would not be pleased if it ever found out.

Oh, knock it off, Tri-face! I kept the Greek letter π (pi) back in Chapter 4 didn't I? The trouble with λ and ν is that they are hard to remember and the stupid ν looks too much like the English letter v that scientists use for velocity (which is very confusing).

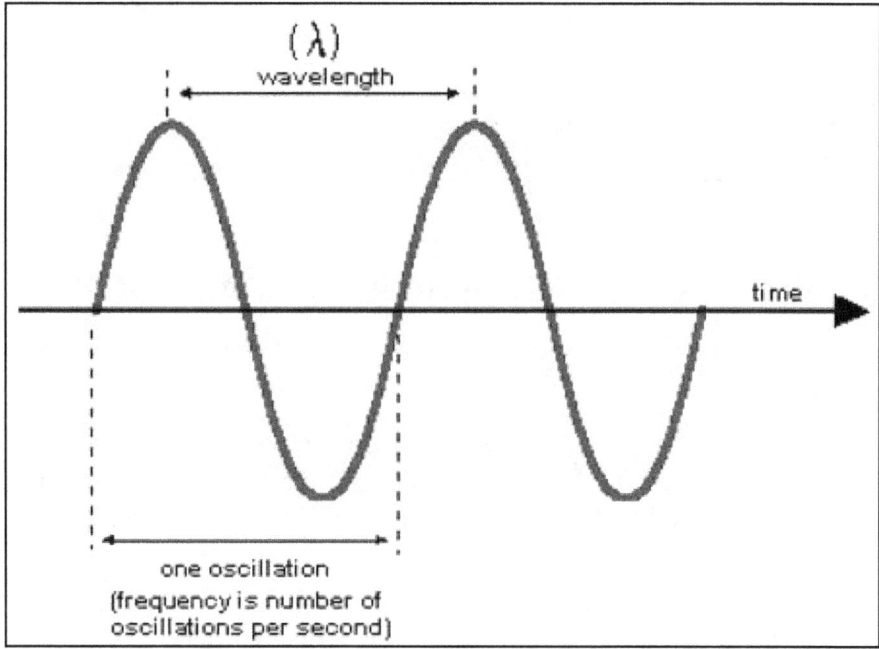

Figure 9-2 Components of a Wave

The Spectrum of Light

Now comes the fun part. The light that we see from the sun is known as **white light** because it is not just at one wavelength. It is made up of a continuous range of wavelengths covering w values from about 400 to 700 nm. Our eyes detect these different wavelengths as different colors.

Isaac Newton demonstrated this by placing a triangular shaped hunk of glass called a **prism** in the path of sunlight with a screen on the other side of the prism (see Figure 9-3). The prism spreads the white sunlight out into separate colors onto the screen. It does this by bending the path of each wavelength a different amount as it passes through the prism. This rainbow of colors is known as a **continuous spectrum** because it contains all wavelengths (with no apparent gaps) within the range of wavelengths coming from the sun. There are other types of spectra (that's plural for spectrum) that are not continuous. These other types play a crucial role in astronomy, as you shall see later in this chapter.

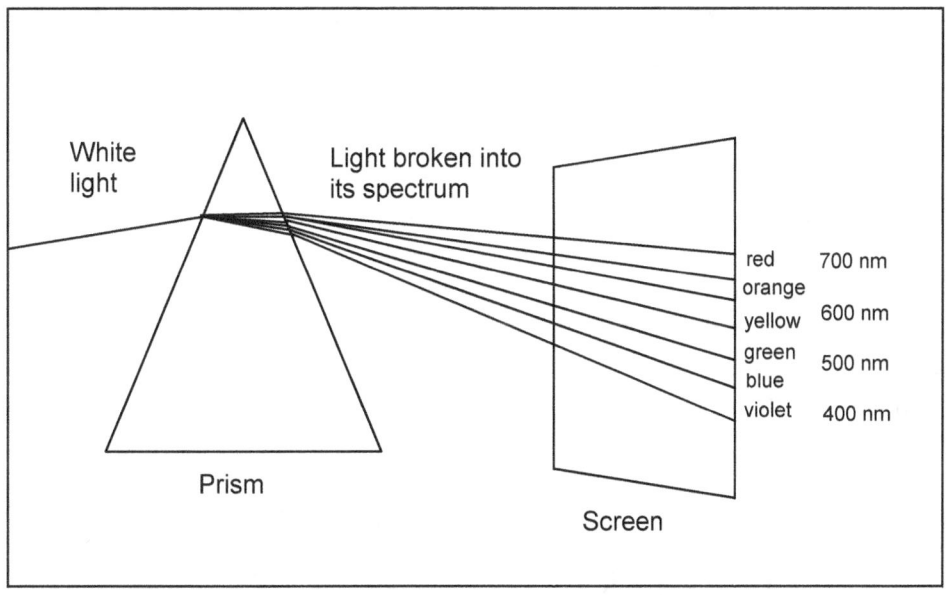

Figure 9-3 Newton's Experiment with White Light

Electromagnetic Spectrum

When we use the word "light", we generally mean visible light, which means all those wavelengths (colors) in Figure 9-3 that the human eye has come to detect after eons of evolution. But this visible light is just a tiny part of a much larger family of radiation, the rest of which our eyes cannot detect — the family of **electromagnetic radiation**. Compared to visible light, the range of wavelengths and frequencies found in the entire electromagnetic spectrum is enormous, as you can see from Figure 9-4.

This radiation is called electromagnetic because its waves are made up of perpendicular, oscillating electric and magnetic fields that ...

HOLD IT, HOLD IT! Good grief, let's not get carried away. This goes way beyond the scope of the book. Like ... keep it simple. And by the way, I got a funny feeling you're gonna be using that word "electromagnetic" a lot, so maybe you should abbreviate it to EM from now on.

OK, OK ... maybe you're right, Tri-face. But let's not get into a big sweat about it. Let me just mention one more thing about eyes.

The reason our eyes have not evolved to detect the rest of the EM spectrum is that evolution focuses on the existing environment in which life finds itself. You know, survival of the fittest and all that. On Earth we find ourselves bathed in sunlight which is predominantly composed of EM radiation in the range of 400 to 700 nm wavelengths. There is little reason for the human eye to develop a detection mechanism for gamma rays, for example, when it is not exposed to those wavelengths.

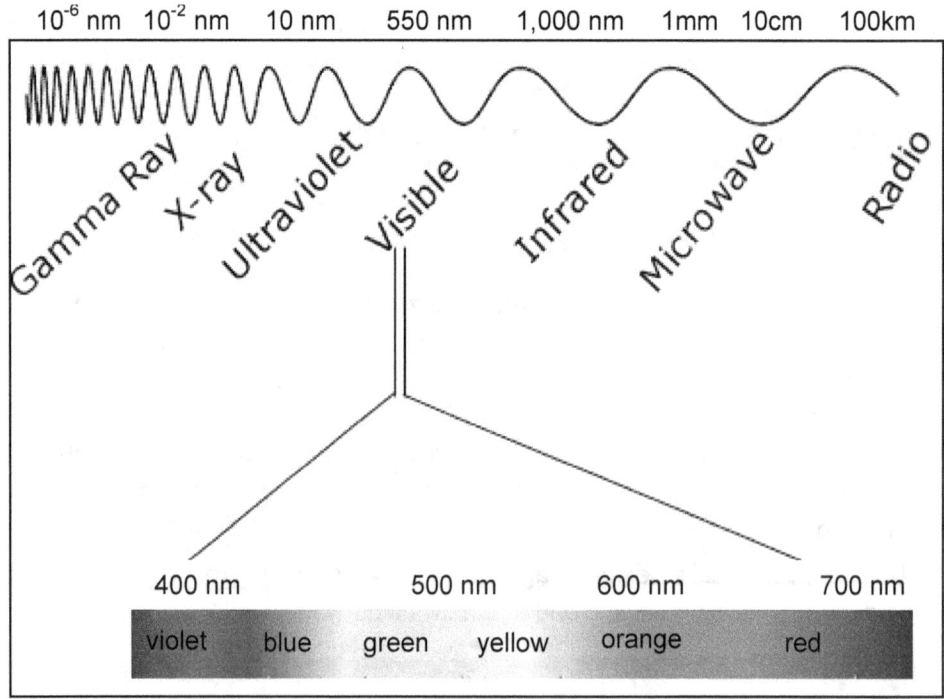

Figure 9-4 Electromagnetic Spectrum

Frequency verses Wavelength

Wavelength values across the whole EM spectrum are shown at the very top of Figure 9-4 and represented by the wavy line beneath these values. They start at the incredibly short wavelengths of 10^{-6} nanometers for gamma rays (forget your everyday ruler — that's a millionth of a billionth of a meter) and extend right up to 100 kilometers or so for the longest radio waves.

The wavy line also symbolizes frequency, starting with the very high frequencies of gamma rays and working down to the very low frequencies of radio waves. So short wavelengths have high frequencies and long wavelengths have low frequencies. This makes sense because the speed of light is constant, so if the wavelength is made shorter, more waves would pass you each second (which means a higher frequency). Mathematically the relationship between wavelength and frequency is dead simple:

$$f = \frac{c}{w}$$

where:

f = frequency (in Hz)
c = speed of light (3×10^8 m/s)
w = wavelength (in meters).

For example, the frequency of a wavelength of 550 nm (the yellow part of the visible light range according to Figure 9-4) would be:

$$f = \frac{c}{w} = \frac{3 \times 10^8}{550 \times 10^{-9}} = 5.45 \times 10^{14} \text{ Hz (or) } 5.45 \times 10^8 \text{ MHz}$$

That's over 500 million megahertz. Compare that to an FM radio broadcast at a frequency of only 100 megahertz. Just for fun, try rearranging this equation to calculate the wavelength for that FM station. (You can do that by multiplying both sides by w and dividing both sides by f to get w on its own.) You should get 3.00 meters for w.

The higher the frequency, the more energetic the radiation and therefore the more harmful it is to the human body. This is why exposure to gamma rays is so deadly. Too much exposure to X-rays can also be dangerous. Some ultraviolet light does reach us from the sun and too much of it can cause severe sunburn. But you knew that.

Energy verses Wavelength

In the early 1900's, German physicist Max Planck derived a really important relationship between the energy of a photon and its wavelength:

$$E = \frac{hc}{w}$$

where:
E = energy of a photon
h = Planck's constant
c = speed of light
w = wavelength

This means a specific energy level for a photon produces a specific wavelength for that photon. We will see the importance of this for astronomers in the next few pages under the heading **Starlight Reveals Chemical Composition**.

Blackbody Radiation

This is the last thing we'll mention about EM radiation before showing you how astronomers apply all this science to their work. Imagine some hypothetical type of object called a blackbody that does not reflect any of the light that falls on it.

A WHAT?

A **blackbody**. A dense object that absorbs all the radiation that falls on it. Unlike dining room tables, Scottish kilts, and people, a blackbody does not *reflect* any radiation. But it does *emit* **thermal radiation** over a continuous range of wavelengths, which may or may not include the visible part of the EM spectrum. It's called "thermal" radiation because the range of wavelengths emitted depends only on the temperature of the blackbody, not its chemical composition. This turns out to be a very important characteristic for astronomers when determining the surface temperature of stars. Why? *Because a star behaves remarkably like a blackbody.*

The higher the temperature of the blackbody, the shorter the *predominant* wavelength of its thermal radiation (and therefore the more energetic it is, according to Max Planck's formula). Also, the higher the temperature, the greater the amount of *total* (all wavelengths) radiation emitted. This is shown in Figure 9-5 for three different blackbody temperatures, each temperature producing its own curve of wavelength versus energy intensity. This energy is simply labeled **intensity** in this graph.

NOTE: For illustrative purposes, the vertical intensity scale has been compressed here so that all three curves can be shown. In reality, the peak intensity for the 12,000 K curve is about 1,000 times greater than that for the 3,000 K curve. Our publisher didn't want to make the page that large.

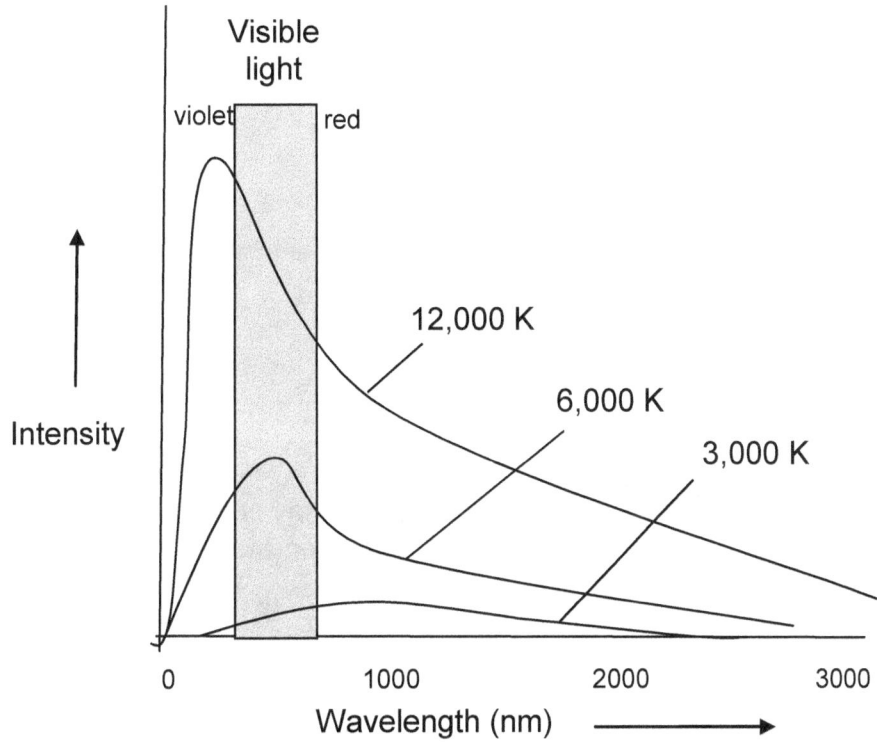

Figure 9-5 Blackbody Curves

Picture a welder heating an iron bar. It absorbs the heat, then emits it as thermal radiation like a blackbody. As the bar becomes hot, it begins to glow deep red. As the temperature rises further, the iron becomes more orange, then bright yellow around 6,000 K. In contrast, a blackbody at 293 K (room temperature) would have so low an intensity at all wavelengths that it wouldn't even show up in Figure 9-5.

But enough of this technical babble about waves, frequencies, white light, Scottish kilts, and blackbodies. Let's see how astronomers make use of this stuff.

Starlight Reveals Surface Temperature

Well, for starters, because stars behave like blackbodies, astronomers can compare a star's emitted light to the whole family of blackbody curves. Figurer 9-5 shows only three members of that family, but you can imagine such a graph including a much larger number of curves representing a much greater number of temperatures.

Figure 9-6 shows how a star's blackbody curve can be matched up with just the right one in the family of blackbody curves to give the surface temperature of that star. In this case, the star is our sun, but the technique works equally well for any star whose EM radiation intensity can be observed and measured at various wavelengths. The technique is called **photometry** and involves attaching a light-sensitive device onto the telescope along with a standard set of color filters to measure the brightness of the different colors (wavelengths) individually.

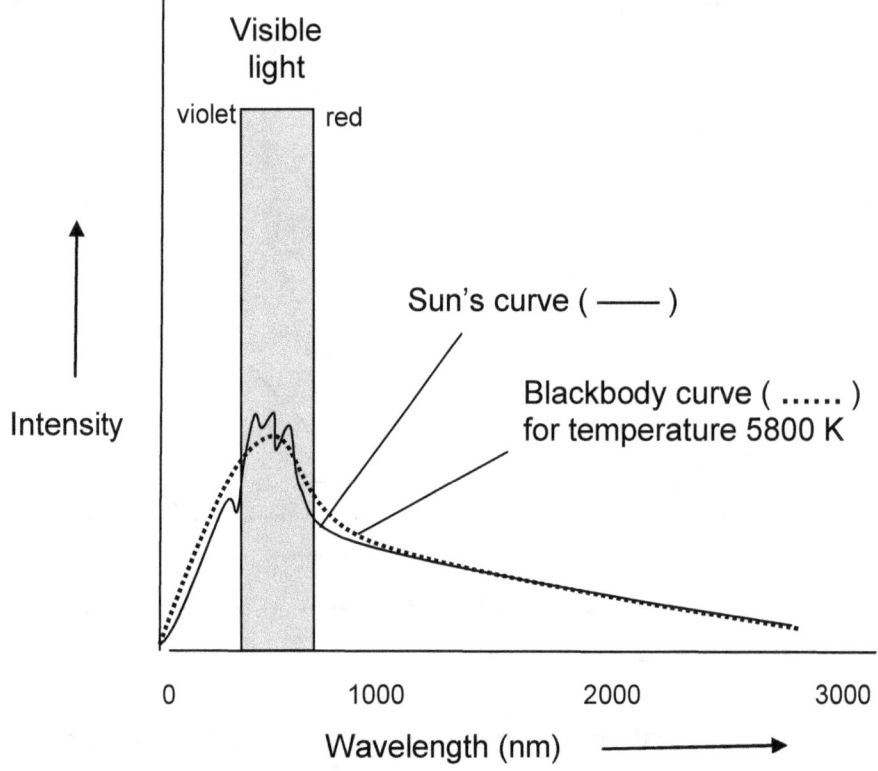

Figure 9-6 The Sun as a Blackbody

Notice that the sun's blackbody curve is slightly jagged because it's not a perfect blackbody. It peaks in the middle of the visible spectrum which just happens to match the 5,800 K blackbody curve. This means the sun's surface temperature must be 5,800 K.

Wien's Law

A handy method of determining a star's surface temperatures that does not involve matching up blackbody curves (as in Figure 9-6) was devised by the German physicist Wilhelm Wien in 1893. He derived a simple formula for blackbody thermal radiation that relates surface temperature to the wavelength of maximum intensity that you see on the blackbody curve. Known as Wien's law, the formula is:

$$W_{max} = \frac{0.0029}{T}$$

where:

W_{max} = wavelength of maximum intensity (in meters)

T = surface temperature (in Kelvins)

As an example of this useful formula, Figure 9-6 shows that the sun's maximum intensity occurs at a wavelength of about 500 nm. Recall that nm is a billionth of meter, so (dividing by 10^9) 500 nm equals 5×10^{-7} m.

If we rearrange the equation by multiplying both sides by T and dividing them by W_{max}, we can calculate the sun's surface temperature as:

$$T = \frac{0.0029}{W_{max}} = \frac{0.0029}{5 \times 10^{-7}} = 5800 \text{ K}$$

Astronomers have made this determination even though the object is 150 million kilometers away.

Auguste Comte would be surprised to learn that no one had to travel to the sun (let alone other stars) with a special thermometer to determine this fact.

As you can see by referring back to Figure 9-5, the EM radiation from many stars that are much hotter or cooler than our sun actually peaks *outside* the visible light range. But astronomers can still see these stars because some of their radiation still falls within the visible range. So a cooler star would look reddish, while a hotter star would look blue or blue-white. Stars really do come in different colors.

Starlight Reveals Chemical Composition

In 1814 the German optician Joseph von Fraunhofer repeated Newton's famous experiment by passing a beam of sunlight through a prism. But he went one step further. He magnified the resulting continuous spectrum enough to discover (to his surprise) that it contained hundreds of fine, dark spectral lines against the rainbow background — unlike the continuous spectrum of a perfect blackbody.

A few decades later, scientists Gustav Kirchhoff and Robert Bunsen discovered that when you heat and vaporize various chemical substances in the lab, the emitted light from the heated substances consists of a bunch of bright spectral lines against a dark background.

So what's with all these funny spectral lines???

This all ties back to our blurb on quantum mechanics on page 21 of Chapter 2 and those crazy, orbit-hopping electrons. Recall that each chemical element has electrons that can jump from orbit to orbit within the atom causing specific changes to the atom's energy level. This, in turn, causes the absorption or emission of photons at those same energy levels. Now according to Max Planck's formula on page 64, photons at specific energy levels will have specific wavelengths. It's these wavelengths that are showing up as the "funny spectral lines" discovered by Fraunhofer and those other guys in white lab coats.

In the case of those dark spectral lines found by Fraunhofer, they are caused by chemical elements in the cooler atmosphere of the sun that absorb specific energy levels of photons from the hotter depths of the sun. In fact, each element can absorb a number of different energy levels specific to that element, because different combinations of orbits are involved in the electron jumps.

When an element absorbs photons in the sun's atmosphere (kicking electrons up to a higher energy level), the corresponding wavelength disappears from the sun's underlying continuous spectrum, leaving a gap at that particular wavelength. This gap shows up as a dark line against the continuous spectrum. The overall result is known as an **absorption line spectrum** because the dark lines indicate absorbed wavelengths.

In the case of those bright spectral lines discovered by Kirchoff and Bunsen, the heated chemical element emits photons when it starts to cool down again (as the electrons drop back to a lower energy level). The resulting radiation consists of a number of different spectral lines that are specific to that element. These lines appear against a black background. The overall result is known as an **emission line spectrum** because the bright lines indicate emitted wavelengths. All this brings us to a number of key points:

- The wavelengths that a given chemical element absorbs in an absorption line spectrum are the same wavelengths that it creates in an emission line spectrum.
- Each chemical element has its own unique set of lines with its specific wavelengths.
- A chemical element can actually be identified by the wavelengths of the spectral lines that it absorbs or emits.

Think of these spectral lines as the fingerprints of the chemical element.

Astronomers, using photometry techniques, can record the spectrum of the object being viewed and accurately measure the wavelength values of the lines within them. What a powerful tool for astronomers!

By analyzing the resulting absorption line spectrum from a star's light and comparing it with the emission line spectrum produced by various elements in the lab, they can determine the chemical composition of individual stars. They know, for example, that the sun's mass is 74% hydrogen, 25% helium, and they know what elements make up the remaining 1% (the so called "metals").

And as Tri-face would point out, they didn't have to travel to the sun (let alone other stars) and scoop up a chemical sample to determine this fact.

Kirchhoff's Laws

The concepts introduced here can be summarized by Kirchhoff's three laws below. Figure 9-7 should help you understand these laws.

Law 1 A hot opaque (non-transparent) body, such as a perfect blackbody or a hot, dense gas produces a continuous spectrum — a complete rainbow of colors without any spectral lines. (See LIGHT PATH A in Figure 9-7.)

Law 2 A hot, transparent gas produces an emission line spectrum — a series of bright spectral lines against a dark background. (See LIGHT PATH B in Figure 9-7.)

Law 3 A cooler, transparent gas in front of a hot source of a continuous spectrum produces an absorption line spectrum — a series of dark spectral lines among the colors of the continuous spectrum. (See LIGHT PATH C in Figure 9-7, where the dotted portion of this path represents the light passing through the cloud of cooler gas.) Furthermore, the dark lines in the absorption line spectrum of a particular gas occur at exactly the same wavelengths as the bright lines in the emission line spectrum of that same gas.

Starlight Reveals Radial (line-of-sight) Velocity

You have probably noticed that the sound of a police car's siren has a high pitch when the car is approaching you (not that you've ever been chased by such a vehicle, of course). And when the same car is moving away from you (thank heavens), it emits a low pitch sound. The high pitch is due to the sound waves being crowded closer together as they approach you, producing a shorter wavelength and hence a higher frequency (or pitch). Conversely, the low pitch is due to the sound waves being stretched out as they move away from you, producing a longer wavelength and hence a lower frequency (or pitch).

The same thing is true for light waves.

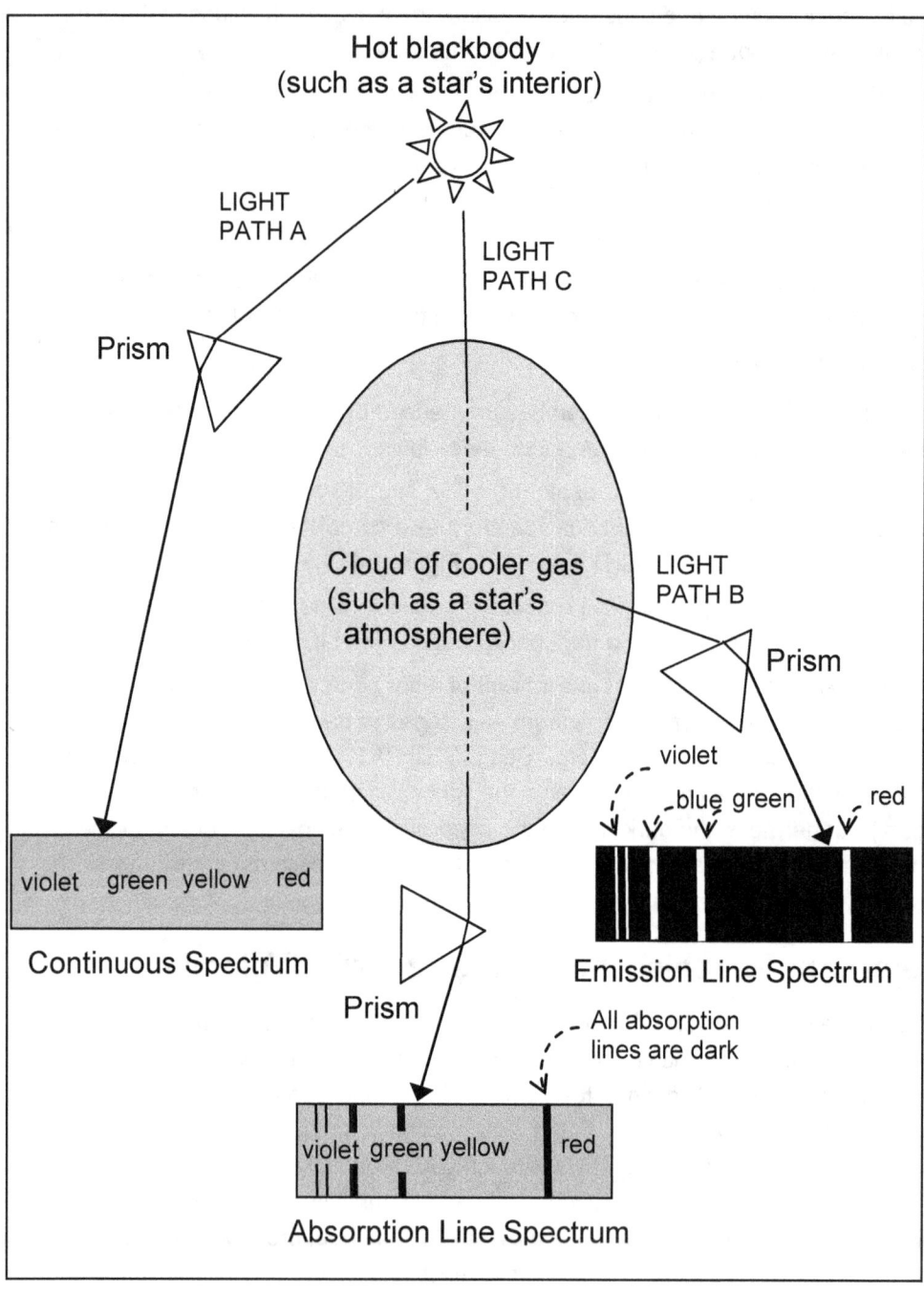

Figure 9-7 Three Kinds of Spectra

Doppler Effect

If a light source is moving toward you (or you toward it — it's all relative), the wavelengths will crowd together more, thus becoming shorter, which — as everybody knows — means they are shifted toward the blue/violet end of the light spectrum. This is called the **blueshift**.

Conversely, if the light source is moving away from you (or you from it), the wavelengths will stretch out more, thus becoming longer, which — as everybody knows — means they are shifted toward the red end of the light spectrum. This is called the **redshift**.

The effect of this relative motion on wavelengths of light is known as the **Doppler effect** or **Doppler shift**, named after Christian Doppler, a mathematics professor who first pointed out this strange effect in 1842. Unlike the effect with sound waves, however, the relative velocity between the light source and the observer has to be much higher than we experience in everyday life before our eyes would notice the effect. After all, you don't expect a red car to look blue as it speeds toward you on the highway.

Don't be like the astronomer who got caught running a red traffic light. He tried to avoid the fine by outlining the Doppler effect to the judge. He explained that because he was approaching the red light at a certain velocity, the color got shifted part way toward the blue end of the spectrum, making the traffic light look green.

Big mistake. It turned out this judge studied physics before going to law school.

Thanks to the astronomer's detailed explanation, the judge did a quick calculation, accepted his green light explanation, dropped the red light charge, then fined him for speeding. The fine was enormous — as you can imagine when you see the sort of speeds involved in the rest of this chapter.

Radial Velocity

The radial velocity of a celestial object is the velocity that takes place along the line of sight as you view the object. In other words, it is the speed with which the object is moving *directly* toward or *directly* away from you. And as we are about to see, there is a direct mathematical relationship between this velocity and the Doppler shift. This means that by accurately measuring the amount of Doppler shift in the wavelengths, astronomers can calculate the corresponding radial velocity. This turns out to be an incredibly useful tool.

 NOTE: This technique does not tell us anything about the sideways motion, only the motion directly along our line of sight to the object.

Doppler Shift Equation

This is pretty straightforward stuff. Certainly not rocket science.

Take w_0 as the wavelength of a particular spectral line from a light source that is not moving. (Think of the zero subscript here as representing zero velocity.) It could be one of the spectra lines from hydrogen or calcium or any other element. You could look up its value in a reference book or determine it in a lab experiment.

Now take w as the wavelength of this same spectral line when measured in the spectrum of a light source that is moving. Because of the Doppler effect, this wavelength will be shifted from the w_0 value. The size of the shift can be written as Δw (delta w), where:

$$\Delta w = w - w_0 \text{ (the change in the wavelength due to the motion)}$$

Scientists and mathematicians just love using the Greek letter delta to denote a change in the value of something. And it will keep Tri-face happy by using it here.

Being a clever mathematician, Doppler proved something extremely useful:

$$\frac{\Delta w}{w_0} = \frac{v}{c}$$

where:

w_0 = wavelength when the source is not moving
Δw = amount of wavelength shift (in same units chosen for w_0)
v = radial velocity of the moving source
c = speed of light (in same units chosen for v)

Think of this relationship as equating two ratios — a wavelength ratio and a velocity ratio. The higher the radial velocity (v) of the light source, the bigger the resulting wavelength shift (Δw).

As you will see in the example that follows, the amount of shift is typically very tiny for stars in our galaxy. Fortunately, wavelength measurements of spectral lines can be done with incredible precision, both in the lab for a light source at rest and on the telescopic spectrograms for a celestial object moving toward or away from us. Would you believe measurements to a thousandth of a billionth of a meter?

Example of the Doppler Shift Equation

Hydrogen has a number of spectral lines, but the prominent one (labeled H_a) has a wavelength of $w_0 = 656.285$ nm as measured in the lab. In the absorption spectrum of the bright star Vega, which is 25.3 light-years away, this line has a wavelength w = 656.255 nm. So its wavelength shift is:

$$\Delta w = w - w_0 = 656.255 \text{ nm} - 656.285 \text{ nm} = -0.030 \text{ nm}$$

The negative value tells us that the light from this star is shifted to shorter wavelengths (blueshift), which means the star must be moving toward us. If it had been moving away from us, this value would have been positive, indicating a shift to longer wavelengths (redshift).

We can now calculate the star's radial velocity by rearranging the Doppler shift equation (multiply both sides by c to get v on its own):

$$v = \frac{c\Delta w}{w_0} = \frac{3 \times 10^5 \times (-0\ 030)}{656.285} = -13.7 \text{ km/s}$$

As you now know, the negative velocity means that this star, which is 25 light-years from Earth, is moving *toward* us at 13.7 km/s. How's that for a nifty scientific accomplishment? Our French philosopher, Auguste Comte, would turn over in his grave if he heard out about this.

OK, dude, now that you have explained the Doppler thing, maybe we should tell the reader about some of the more profound discoveries in which the Doppler effect has played a role.

OK, but what do you mean "we"? I'm the guy doing most of the work here.

The Doppler Effect Marches On

This powerful technique of measuring velocities of distant objects along their line of sight has had far reaching consequences in the advancement of cosmology.

For example, astronomers have used the Doppler technique to plot the motions of many stars in our Milky Way galaxy, which has led to an understanding of how the galaxy is rotating. And when they figured out the speed of that rotation and compared it to what you get from Newton's universal law of gravitation and his other laws of motion, they realized that something was wrong. Something was amiss. This story unfolds in Chapter 15 with the concept of dark matter.

Another example is in the determination of the recessional velocities of distant galaxies. Unlike the stellar velocities found inside our home galaxy, these galactic velocities are measured in thousands of kilometers per second and so exhibit very large redshifts. This story unfolds in Chapter 11 with the concept of an expanding universe and again in Chapter 15 with the concept of dark energy.

 NOTE: In the last chapter, we mentioned that Newton's universal law of gravitation, while very useful for most purposes, had to give way to Einstein's general theory of relativity when dealing with very strong gravitational fields. The same idea applies to the Doppler shift equation when dealing with very high velocities. It needs to be modified. We'll pick this story up in Chapter 11 where you too will get to practice some relativity.

The Doppler technique has also been used to discover planets around other stars by measuring the wobbly motion of these stars — a motion caused by the gravitational tug of the unseen planets. Because this motion is incredibly slow by celestial standards, the resulting wavelength shifts are very difficult to measure. Chapter 17 touches on this topic in the search for alien life.

Christian Doppler would have been astonished to see that the simple equation he derived in the mid 1800's has played such a major role in cosmology throughout the 1900's and continues to do so today.

Chapter 10 —
The Distance Ladder

This chapter is all about determining distances, so you may be wondering about the "ladder" part. Let me explain before Tri-face gets on my case. The purpose of any rung of a ladder is to get to the next higher rung until you reach the height you want. It would be difficult to get to the third rung if the bottom two were missing.

And so it is with our symbolic ladder for determining astronomical distances. The first rung uses a simple technique that works for nearby stars. The second rung makes use of the results from the first rung, combined with a different technique, to determine distances to stars farther away (even to nearby galaxies). Each subsequent rung makes use of the results from the previous rung, combined with yet another technique, to extend these distance determinations even farther out. At the top of the ladder, astronomers are able to determine distances to remote galaxies measured in billions of light-years.

It turns out that the most widely used rungs on the ladder are the first two and the last two. So, if Tri-face doesn't mind, we'll skip a few rungs in the middle.

The First Rung (stellar parallax)

The technique on this rung only works for stars close to home. It's based on a phenomenon know as **parallax** — something we all experience every day. When you view an object close by while walking or driving past it, it changes its position with respect to the more distant background. You can demonstrate this concept for yourself. Just stretch your arm out in front of you, hold up one finger, and view that finger with one eye closed, then the other. Go ahead. Give it a try right now.

You might wanna make sure no one is watching while you do this.

Notice how your view of your finger shifts against the background as you switch from one eye to the other. That shift is called **angular displacement** and is caused by observing an object from two different positions (in this case, your two eyes). The distance between your eyes creates this effect. This distance between one position of observation and another is known as the **baseline** when we talk about parallax.

The same principle applies when you view a planet or star. As you change your viewing position, the planet or star shifts against the background of far more distant stars. Of course, you now need a much longer baseline than provided by your two eyes. You could, for example, view Mars or Jupiter from two different locations on

Earth. If the two locations were on the equator on opposite sides of the globe, you would have a baseline equal to Earth's diameter (12,756 km). But alas, even this baseline is too short to detect the angular displacement of stars. They're just too far away.

For stars we need yet a larger baseline. Fortunately, we have one — the diameter of Earth's annual orbit around the sun, which is two AU or about 300 million kilometers. You just have to wait six months between observations.

Now for the fun part. Figure 10-1 sets the stage for calculating stellar distances based on parallax. It shows the lines of sight when observing a nearby star at opposite times of the year. Obviously, this diagram is not drawn to scale.

 Don't panic. Despite the weird terminology coming up, this is a really simple calculation.

The measured angle between these two lines of sight is the angular displacement against the background stars. The parallax of the star is denoted by the letter p in the diagram and is defined as half of this angular displacement. The objective here is to determine the distance to the star, as represented by the letter d. Here's the parallax formula:

$$d = \frac{1}{p}$$

where:

d = distance to the star (in parsecs)

p = parallax angle of the star (in arcseconds).

How's that for a simple equation? The reason it's so simple is because of the new units of measurement involved.

Let's take the easier unit first: **arcseconds**. Angles are measured in degrees and there are 360 degrees in a complete circle. What you might not know is that astronomers need to measure really tiny angular displacements, so they like to subdivide one degree into 60 **arcminutes** (abbreviated arcmin). Not being satisfied with that, they subdivide each arcminute into (you guessed it) 60 **arcseconds** (abbreviated arcsec). So there are 60 x 60 or 3,600 arcsecs in just one degree. One arcsecond is a very tiny angle.

Now for this new distance unit, **parsecs** (abbreviated pc). The name is a contraction of the words **pa**rallax and arc**sec**ond. Not being content with astronomical units (AU), and light-years (ly), astronomers like to use parsecs as well. Why? Because it provides the simple parallax formula introduced above. My dictionary of astronomical terms defines a parsec as *the distance at which the semimajor axis of the Earth's orbit subtends an angle of one arcsecond.*

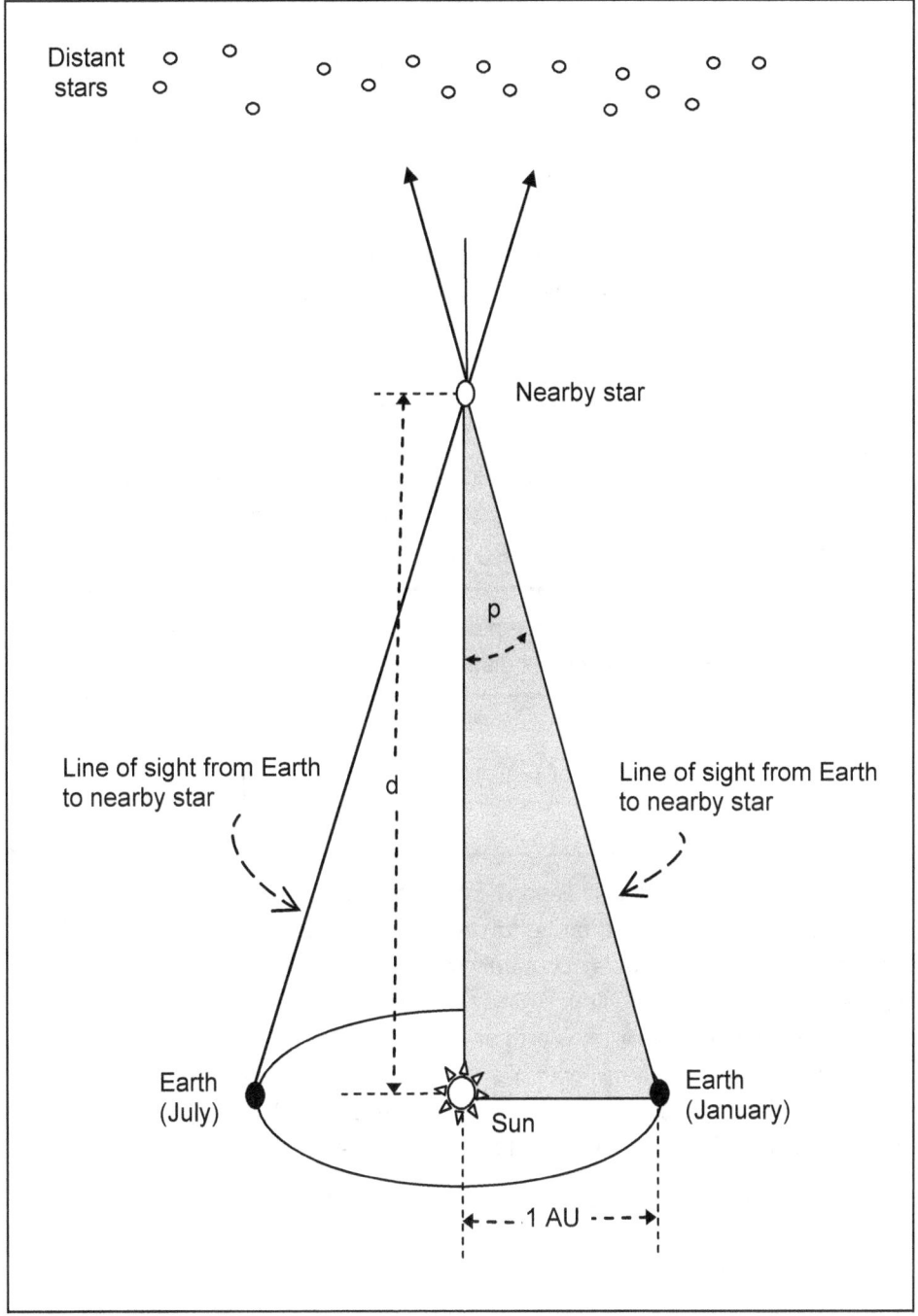

Figure 10-1 Stellar Parallax

Uh? Could you please repeat that in plain English.

I'll try. In Figure 10-1 the sun-to-Earth (January) distance represents the "semimajor axis of the Earth's orbit" (a component of an ellipse, similar to the radius of a circle). The angle p represents the angle that this axis "subtends". *When the size of that angle is set to one arcsec, then the sun-to-nearby star distance (d) is* <u>*DEFINED*</u> *to be one parsec.*

NOTE: One parsec turns out to be 3.260 light-years, as calculated from the gray-shaded right-angled triangle in Figure 10-1 using a branch of mathematics called trigonometry. There are trigonometric function keys on your scientific calculator that make this calculation really easy — but we're not going there today.

OK, so if someone asks you how far away the nearest star, Proxima Centauri, is, you can say 4.2 light-years or (dividing by 3.26) you can say 1.3 parsecs.

Right on! Now let's put this simple parallax formula to work. If a star has a measured parallax of 0.545 arsec, its distance is:

$$d = \frac{1}{P} = \frac{1}{0.545} = 1.83 \text{ pc} \ \text{(or)} \ 1.83 \text{ pc} \times 3.26 \text{ ly per pc} = 5.98 \text{ ly}$$

How's that for a simple calculation? You can see from Figure 10-1, or from the parallax formula itself, that the farther away a star is, the smaller its parallax. Ground based telescopes have trouble accurately measuring really small parallaxes because of atmospheric turbulence. Observations from Earth-orbiting satellites, however, can measure parallaxes with an accuracy of 0.001 arcsec. That's considerably less than one millionth of a degree. This enables astronomers to determine stellar distances as far out as ... well (using the formula)... 1 divided by 0.001. That's 1,000 parsecs or more than 3,000 light-years.

But that's the maximum distance we can get out of the first rung of the ladder. When you consider that the center of our own galaxy is some 26,000 light-years away and cosmologists want to determine the distances of remote galaxies, you can see that 3,000 light-years is nothing in the scheme of things. Check out Table 3.1 again on page 26.

The Second Rung (Cepheid variable stars)

As you just saw, the parallax method is based on geometric measurements. The technique used on the second rung of the distance ladders is entirely different. It's based on three quite separate concepts:

- the inverse-square law
- standard candles
- Cepheid variable stars.

Oh sure. As if everybody knows what you mean by the "inverse-square law" and ... WHAT kind of stars???

Keep your shirt on, Tri-face. You're about to find out.

NOTE: You need to be really focused here. Just think of this as a carefree academic exercise (no exam!). Try to grasp each concept in turn, even if you have to read it twice. Don't worry about how it fits in with the final goal of determining distances. It should all come together for you in the end. If you miss a concept or two, it's no big deal. Just try to get a feel for what a little science can do for us here.

The Inverse-Square Law

Imagine placing a 60 watt light bulb at a distance of one meter away and observing its brightness, then repeating the process at two meters, then again at three. If you could somehow measure the intensity of the light reaching your eyes, you would discover that doubling the distance reduces the brightness to a quarter of your original observation (not a half). And tripling the distance reduces the brightness to a ninth (not a third).This is what the inverse-square law is all about.

Hey, dude, you forgot to mention that before you start the experiment, you should turn the dumb light bulb ON.

OK, let's just ignore Tri-face for now and define the **inverse-square law**:

The apparent brightness of light that an observer can see or measure is inversely proportional to the square of the observer's distance from the light source.

This law is illustrated in Figure 10-2. It's a two dimensional drawing, so you have to imagine the light source being centered within three concentric spheres. The unit of distance used here is the meter, but it could just as easily be light-years or anything else.

The total amount of energy emitted by a light source each second is known as its **luminosity** and is expressed in watts (abbreviated W). (For the curious reader, a watt equals one joule per second where joule is a unit of energy).

Figure 10-2 highlights a small portion of this luminosity falling on one square meter of the first sphere. Notice how this portion gets spread out to four square meters on the second sphere, then nine square meters on the third. This means the amount of light falling on each square meter of these spheres gets less the farther out you go. The light received at each sphere is known as the source's **apparent brightness** and is measured in watts per square meter (abbreviated W/m^2).

So the apparent brightness of the source, as measured on one of the spheres, would be the luminosity of the source divided by the sphere's entire surface area over which this total luminosity is spread. The surface area of a sphere with radius d is $4\pi d^2$, so mathematically, the inverse-square law in astronomical terms looks like this:

$$b = \frac{L}{4\pi d^2}$$

where:

b = star's apparent brightness (in W/m^2)

L = star's luminosity (in W)

d = star's distance from Earth (in m)

This is another simple, but incredibly powerful, formula for astronomers. It contains three variables: b, L, and d. If you know the value of any two of them, you can calculate the value of the third one. (Isn't that what algebra is all about?)

But guess what. Astronomers can always measure the apparent brightness. So if they know the star's luminosity, they can calculate its distance. Or — and this turns out to be equally important for the distance ladder — if they know the distance, they can calculate the luminosity. Neat stuff.

Just for kicks, let's calculate the luminosity of the sun. Scientists have measured our sun's apparent brightness here on Earth at about 1,380 W/m^2 and we know its distance is one AU or 1.50×10^{11}m. So rearranging the inverse-square law equation by multiplying both sides by $4\pi d^2$ gives:

$$L = 4\pi d^2 b = 4 \times 3.14 \times (1.50 \times 10^{11})^2 \times 1380 = 3.90 \times 10^{26} \text{ W}$$

We're not talking about a 60 W light bulb here! Thanks to the inverse-square law, we just figured out how much energy is pouring out of an object 150 million kilometers away.

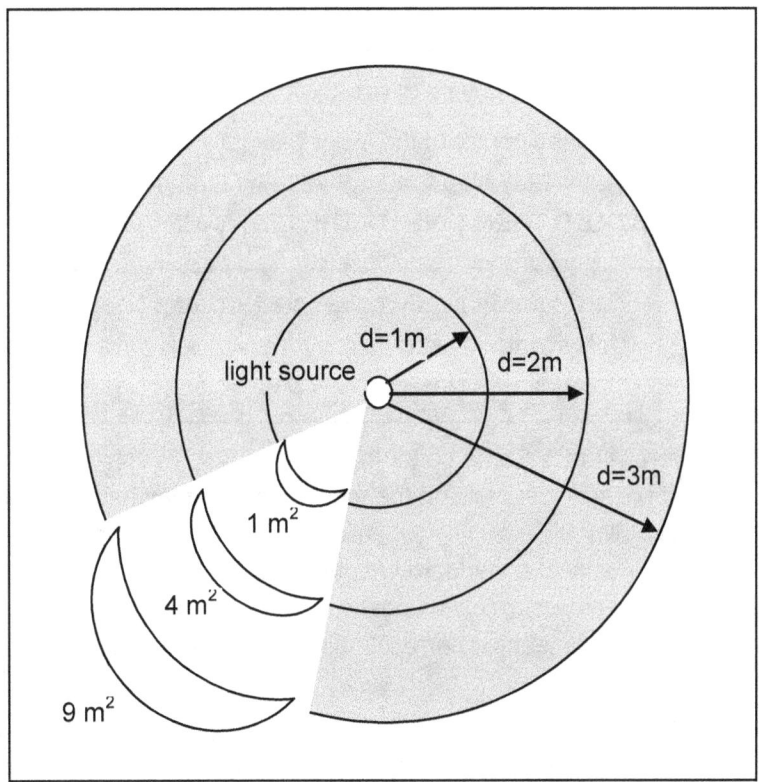

Figure 10-2 The Inverse-Square Law

Standard Candles

Imagine yourself looking at a light source (yet again). If you had the right equipment, you could determine its apparent brightness (b) by doing a photometric measurement. Now, if you happened to know its luminosity (L), you could calculate its distance (d) by rearranging the inverse-square law formula once again. Multiplying both sides by d^2 and dividing them by b gives us:

$$d^2 = \frac{L}{4\pi b} \text{ (or) } d = \text{ the square root of } \frac{L}{4\pi b}$$

In the case of the light bulb, we do know the luminosity. It's 60 watts. So we could calculate its distance. Now imagine some special kind of star that — like the light bulb — had a *known* luminosity. It would be like having a **standard candle** in the night sky — a star with some *standard* luminosity. Such a star would serve as a **distance indicator**, because you could always calculate its distance from the inverse-square law.

Now here comes the really cool part. If we could find and identify such a star in the night sky, we could calculate its distance using the inverse square law ...

...NO MATTER HOW FAR AWAY IT WAS.

It could be halfway to the center of the Milky Way galaxy. It could be many thousands of times farther away than that — in another galaxy ...

... AND THAT WOULD THEN GIVE US THE DISTANCE TO THAT GALAXY.

That sounds really awesome. But are there actually such standard stellar candles? Such convenient distance indicators?

You bet! Several, in fact. Some better than others. But I'm only going to talk about the two that cosmologists rely on the most, because they have a lot of confidence in their ability to determine (or calibrate) their luminosities. After all, determining the luminosity of a standard candle is really what makes it a standard candle. And doing this as accurately as possible is very important, because any error in the determination means a corresponding error in any distance calculations based on that standard candle.

The standard candles that astronomers use (including the two about to be presented) all have the following necessary characteristics to be a useful standard:

- They are very bright and therefore visible out to great distances.
- They are easy to identify by their spectrum or their light curve (see below).
- They have luminosities that astronomers have been able to determine reliably.
- They are relatively common and therefore available for determining distances to lots of galaxies.

Cepheid Variables

The standard candle (or distance indicator) that astronomers use on the second rung of the distance ladder is know as a **Cepheid variable**. These are unstable dying stars that go through regular periods of swelling and shrinking during their death throes. These pulsations cause corresponding changes in the star's luminosity and (therefore) apparent brightness. This periodic variation ranges from a few days to a few months, depending on the individual Cepheid. By measuring its apparent brightness over a number of pulsating cycles, astronomers can plot its **light curve**.

Figure 10-3 shows such a plot over several days for a Cepheid named Delta Cephei. In this plot, apparent brightness is not expressed in W/m² but in **apparent magnitudes**, a rather weird brightness scale that works backwards — the brighter the object, the lower its magnitude value. The use of this scale, however, has no affect on our discussion here.

NOTE: Astronomy is an ancient science. The magnitude scale originated in Greece in the second century B.C., but has since been refined. We'll spare you the details.

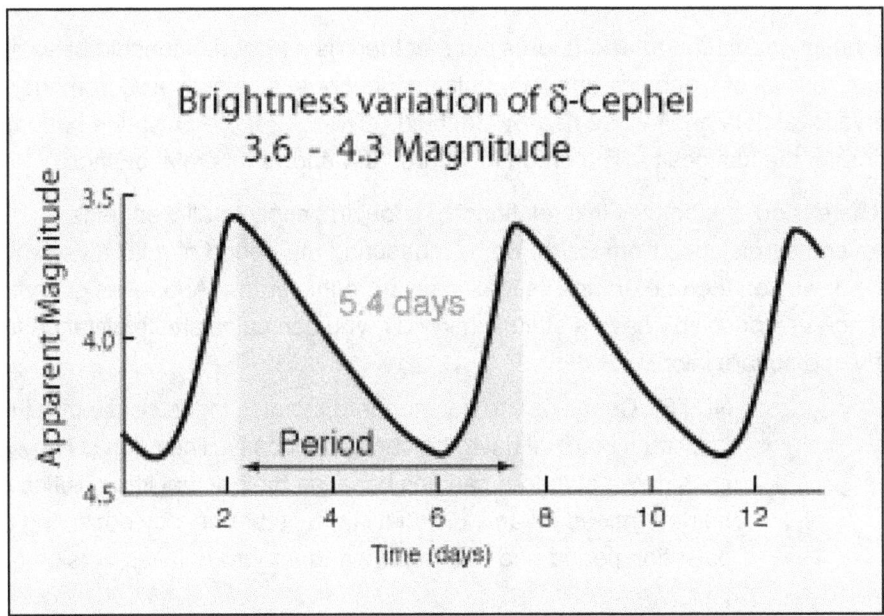

Figure 10-3 Light Curve for Delta (δ) Cephei

At this point you may be wondering — why are Cepheid variables considered to be standard candles? Don't we need to know their luminosities if they are going to play this role? The answer is … yes, we do. And this is where the distance ladder concept comes in.

Astronomers have measured the apparent brightness of many Cepheids that happen to be nearby. By nearby, we mean close enough to allow astronomers to determine their distances by stellar parallax (the technique of the first rung of the ladder). BINGO!! Knowing their distance and apparent brightness, astronomers can calculate the luminosities of these nearby Cepheids by rearranging the inverse-square law as we did for the sun on page 80:

$$L = 4\pi bd^2$$

What astronomers are doing here is *calibrating* Cepheid luminosities at first rung distances so that these types of stars can then be used as standard candles when they are identified farther out at second rung distances. This is where one rung of the distance ladder supports the next higher one.

I don't get it. If individual Cepheids have different luminosities, how do you know the luminosity of an individual Cepheid that is far away at second rung distances?

I thought you'd never ask. It turns out that there is a neat relationship between the luminosities of Cepheids and their pulsating periods, as discovered from the observational data of all those nearby Cepheids. The longer the pulsating period, the greater the luminosity. The graph in Figure 10-4 shows this relationship.

There is no reason why this relationship shouldn't apply to all Cepheids, whatever their distance from Earth. So by measuring the period of a far away Cepheid, we can then determine its luminosity from this graph. And — as everybody now knows — once you have a star's luminosity, you can calculate its distance from the inverse-square law.

NOTE: Cepheids are not standard candles in the sense that these individual stars all have the same luminosity. They don't. They are considered standard candles because their actual luminosities can be determined on an individual star by star basis by observing their pulsation period and then referring to a more detailed version of Figure 10-4.

There are a couple of things about this graph that require some explanation:

Solar Units — Astronomers often express stellar properties like mass, luminosity, and radius in terms of our sun. (Recall the 25-M⊙ star in Table 5-1 on page 36.) In the case of Figure 10-4, the luminosity axis (L⊙) is not expressed in watts, but in solar luminosities. As the dotted lines show, a Cepheid with a 10 day pulsation period would have a luminosity of about 3,000 L⊙, which means 3,000 times the luminosity of our sun. (These guys are big.) Admittedly, it's hard to determine the 3,000 L⊙ value here because this particular graph hasn't been drawn large enough to show grid lines and more values along the X and Y axis. But you get the idea.

Logarithmic Scales — The X and Y axis are logarithmic scales in which values go up by powers of ten. It's just a way of covering a large range of values on one graph.

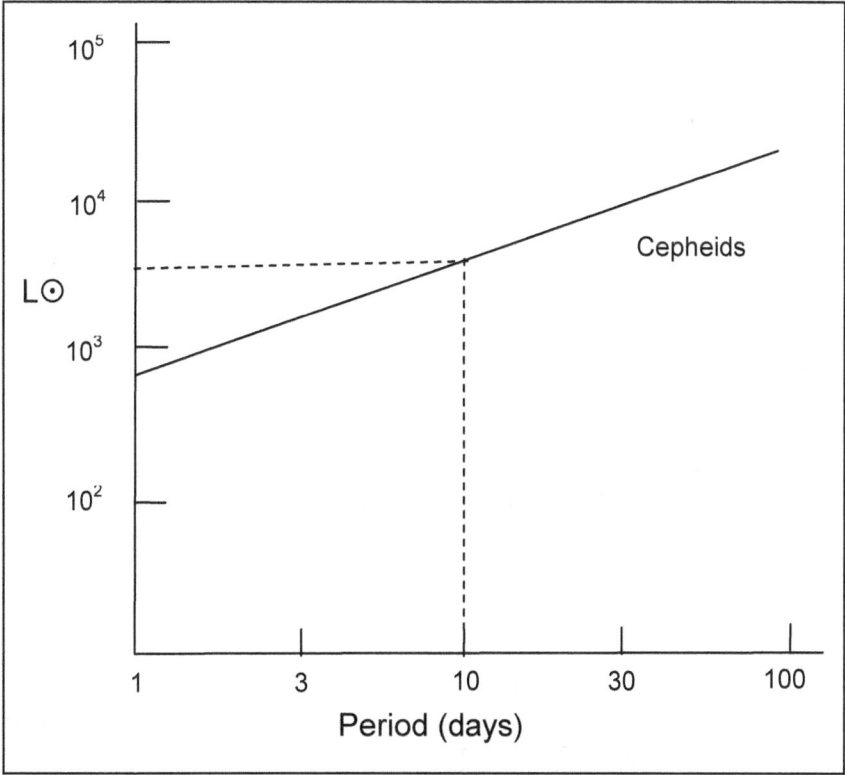

Figure 10-4 Period-Luminosity Relationship for Cepheids

In summary, here are the steps that you, as an astronomer armed with your handy telescope, would typically go through to determine the distance of a galaxy:

1. Search for a Cepheid variable in that galaxy until you find one.
2. Observe it long enough to determine its pulsation period in days.
3. From step 2 determine its average luminosity (L) from a source like Figure 10-4.
4. Measure its average apparent brightness (b) over a few pulsation periods.
5. Calculate its distance using:

$$d^2 = \frac{L}{4\pi b} \quad (or) \quad d = \text{the square root of } \frac{L}{4\pi b}$$

Wait a minute. This formula uses watts for its unit of luminosity (L). What are we supposed to do with a value like 3,000 solar-luminosities?

Good grief, Tri-face, haven't you been paying attention? A few pages back we calculated the luminosity of the sun and found it to be 3.90 x 10²⁶ W. So that's the value of one L⊙. Just multiply that figure by 3,000 to get your watt units. These large figures are going to give you an awfully big number for the star's distance, but that's because this distance is in meters in this formula. You would obviously want to convert it to light-years or parsecs before reporting it to anyone. Come on now, Tri-face, don't tell me that _YOU_ don't have access to a scientific calculator.

I need to see an example.

OK, let's calculate the distance of galaxy IC 4182 by using data from a Cepheid variable found within it. This Cepheid turns out to have a period of 42 days and an average apparent magnitude of 22 according to the literature. So what's its distance and hence the distance to galaxy IC 4182?

If we had a good blown up version of Figure 10-4, we would see that 42 days gives us an average luminosity of 33,000 L⊙. (This Cepheid is monstrous!) So to use our formula,

$$d^2 = \frac{L}{4\pi b} \quad (or) \quad d = \text{the square root of } \frac{L}{4\pi b}$$

we need to multiply 33,000 L⊙ by the sun's luminosity of 3.90 x 10²⁶ W to convert L to watts. That works out to 1.29 x 10³¹ W.

Now — without dragging you through the awkward conversion — the apparent magnitude of 22 is equivalent to 4.36 x 10⁻¹⁷ W/m² which is the unit we need for b in the formula. To get an idea how terribly faint this Cepheid is, compare this tiny figure with the sun's apparent brightness of 1,380 W/m². Because of the inverse-square law, even luminous stellar giants become really faint at large distances.

Now that we have the right units for L and b, we can calculate the distance (d):

$$d = \text{square root of } \frac{L}{4\pi b} = \text{square root of } \frac{1.29 \times 10^{31}}{4\pi \times 4.36 \times 10^{-17}}$$

$$= 1.53 \times 10^{23} \text{ m}$$

That's a lot of meters, so let's divide our answer by the number of meters in a light-year (see Reference Data section) to get the distance in light-years:

$$\frac{1.53 \times 10^{23}}{9.46 \times 10^{15}} = 1.62 \times 10^{7} \text{ ly (or) about 16 million light-years}$$

Cepheids are very massive stars, with luminosities measured in thousands of solar luminosities. As a result, you can observe these standard candles as far out as 200 million light-years. That's a long way beyond the Andromeda galaxy. But farther than that, Cepheids become too dim to see, even with the Hubble Space Telescope. So that's the limit for the second rung of the distance ladder.

A more luminous standard candle is needed determine the distance of remote galaxies billions of light-years away. So up the ladder we go …

The Second Highest Rung (Type Ia Supernovae)

 NOTE: The middle rungs that we are ignoring in this chapter are based on other standard candles that are progressively more luminous than Cepheids. They include really massive dying stars (known as **supergiants**) and even globular clusters (when the distances are too great to distinguish individual stars).

Astronomers — especially cosmologists — use **Type Ia supernovae** as standard candles for determining distances of very remote galaxies because they are more luminous than any of the other standard candles. In general, supernovae are gigantic explosions of massive dying stars, a subject covered in Chapter 14. Type Ia is a special kind of supernova that produces a very sudden and very powerful thermonuclear explosion, creating a peak brightness that — for a few days — can be as bright as the entire galaxy in which it occurs! The explosion can be seen at distances measured in billions of light-years.

The useful thing about Type Ia is that its peak luminosity is the same for all explosions of that type. So if astronomers can somehow determine the value of this peak luminosity, then BINGO — we have an incredibly useful distance indicator for really remote galaxies.

Type Ia explosions, however, are rare. It's estimated that our Milky Way galaxy averages about one every century. Nevertheless, because there are billions of galaxies in the cosmos, astronomers can find lots of these events with today's powerful telescopes. A key point here is how these explosions fit in with the concept of the distance ladder:

- Many Type Ia supernovae have been observed within galaxies that are near enough to have observable Cepheid variables.

- Thanks to those Cepheids, we can determine the distances to those galaxies and therefore the distances to the supernovae explosions within them.

- Knowing those distances, we can then (you guessed it) calibrate the peak luminosity of these explosions using the inverse-square law — *just as Cepheid luminosities are calibrated from their distances as determined by parallax.*

Figure 10-5 shows the light curve associated with this type of supernova. In about one day the explosion reaches a maximum brightness of about three billion solar luminosities (3×10^9 L⊙) before gradually declining over a period of months. No wonder astronomers can spot them so far away.

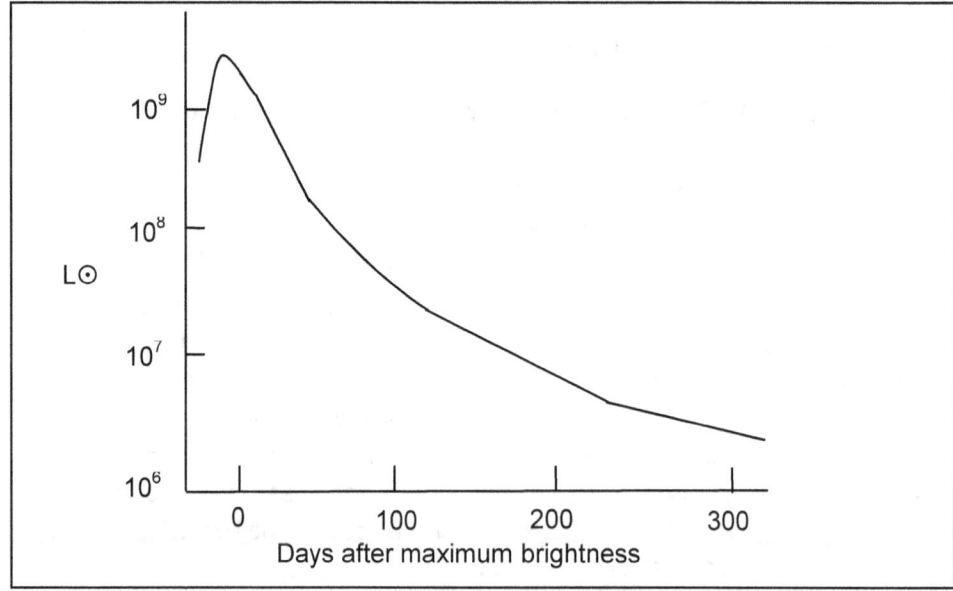

Figure 10-5 Type Ia Supernova Light Curve

In 1999, the Canada-France-Hawaii telescope found one of these supernovae (SN 1999fv) at a distance of nine billion light-years. Imagine having a standard candle available to make distance determinations that far away.

 And you now know how this works, right? You measure the supernova's apparent brightness (b), convert its luminosity of 3×10^9 L⊙ to watts (L), then use the inverse-square law to calculate its distance (d).

But even these bright beacons become too faint to determine some of the extremely remote galaxies that are being observed today, some reportedly as far away as 13 billion light-years. For this, astronomers have to abandon the standard candle approach and fall back on the Hubble law.

The Highest Rung (the Hubble Law)

Remember this law back in Chapter 6?

$$v = H_0 d$$

where:
v = recessional velocity of a galaxy
H_0 = the Hubble constant
d = distance to the galaxy.

You can use it to determine the distances of a galaxy if you know its recessional velocity. All you have to do is rearrange the equation by dividing both sides by H_0. (Mr. Hubble won't mind, he's dead):

$$d = \frac{v}{H_0}$$

This final rung of the ladder has something in common with the first rung, namely…

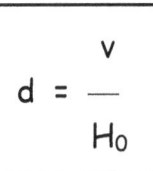 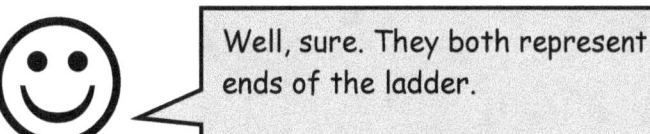 Well, sure. They both represent ends of the ladder.

… well, yeah, but I was thinking of something a little more significant. Namely, that the parallax technique and the Hubble law are the only two techniques on the ladder that do not use standard candles.

This final rung does, however, apply the concept of calibration by making use of the rung below it. Astronomers have to calibrate the constant H_0 because they obviously need to know its value in order to calculate the distance (d).

This is a good-news bad-news story.

The bad news is that cosmologists have had trouble agreeing on the value of H_0 ever since Hubble came up with his famous law. The good news, however, is that they are finally getting close to an agreed value of about 74. We'll discuss the units for this law in the next chapter.

They accomplished this calibration by using the previous rung of the ladder to get the distances (d) of galaxies that contain observable Type Ia supernovae. They then determined the recessional velocities (v) of those same galaxies by measuring their Doppler shift (see Chapter 9). With d and v values now established for a number of galaxies, they calculated the Hubble constant in each case by rearranging the Hubble law equation once again:

$$H_0 = \frac{v}{d}$$

They then averaged the results for H_0 from those galaxies. With an agreed value for H_0, cosmologists now use the Hubble law as a distance indictor for the most remote galaxies they can spot. We'll provide an example of this in the next chapter when we discuss this law more fully.

Chapter 11 —
The Hubble Law

Back in Chapter 6 when discussing the age of the universe, we introduced the Hubble law — a very simple mathematical statement about the expansion of the universe. This law popped up again in Chapter 10 as a method of determining very remote cosmic distances. Obviously, this law is of great interest to cosmologists.

In this chapter, we will describe the units of measurement used in the Hubble law and illustrate its role in determining (1) the distances of remote galaxies and (2) the age of the universe.

The Law ... Once Again

But this time with units:

$$v = H_0 d$$

where:

$v =$ recessional velocity of a galaxy (in km/s)
$H_0 =$ the Hubble constant (in km/s/Mpc)
$d =$ distance to the galaxy (in Mpc).

Mpc stands for a million parsecs. We're talking really great distances here, so the law uses big units for distance. (Recall from Chapter 10 that one parsec is equal to 3.26 light-years.) The constant, H_0, needs some explanation, because the unit km/s/Mpc looks a little confusing. The rearrangement of the equation below shows what H_0 is equal to:

$$H_0 = \frac{v}{d}$$

The v represents the rate of expansion of the universe (in km/s) at some specific distance(d) from Earth. The farther that distance, the faster the expansion. In other words, the Hubble constant is the rate at which the expansion velocity of the universe changes with distance. If we use the generally accepted value of 74 for H_0, this means that the universe is expanding at a rate of 74 km/s for each million parsecs (Mpc) of distance from Earth.

So, for example, a galaxy 100 Mpc from Earth would be receding from us at 7,400 km/s, while a galaxy at 200 Mpc(twice as far away) would be receding at 14,800 km/s (twice as fast) — as calculated by the Hubble law:

$$v = H_0 d = 74 \times 100 = 7{,}400 \text{ km/s} \quad \text{and}$$
$$v = H_0 d = 74 \times 200 = 14{,}800 \text{ km/s}$$

I wonder how fast a galaxy 13 billion light-years away would be receding from us? To make this calculation, we need to convert the light-year distance to Mpc by dividing it by 3.26 (to get pc) and then dividing again by 1,000,000 to get Mpc:

$$v = H_0 d = 74 \times \frac{13 \times 10^9}{3.26 \times 10^6} = 295{,}092 \text{ km/s}$$

WOW! The speed of light is 300,000 km/s, so that galaxy is moving away from us at almost the speed of light. This is why, back in Chapter 7, when we first discussed the expansion of the universe, we mentioned that galaxies beyond our observable universe would be receding faster than the speed of light. We pointed out that this was possible because it is really the expansion of space itself (carrying the galaxies along with it) that is exceeding the speed of light.

Figure 11-1 is a 1996 plot of actual data from galaxies whose distances and recessional velocities have been determined out to almost 500 Mpc (or 1.6 billion ly). It illustrates the linear (straight line) relationship of the Hubble law.

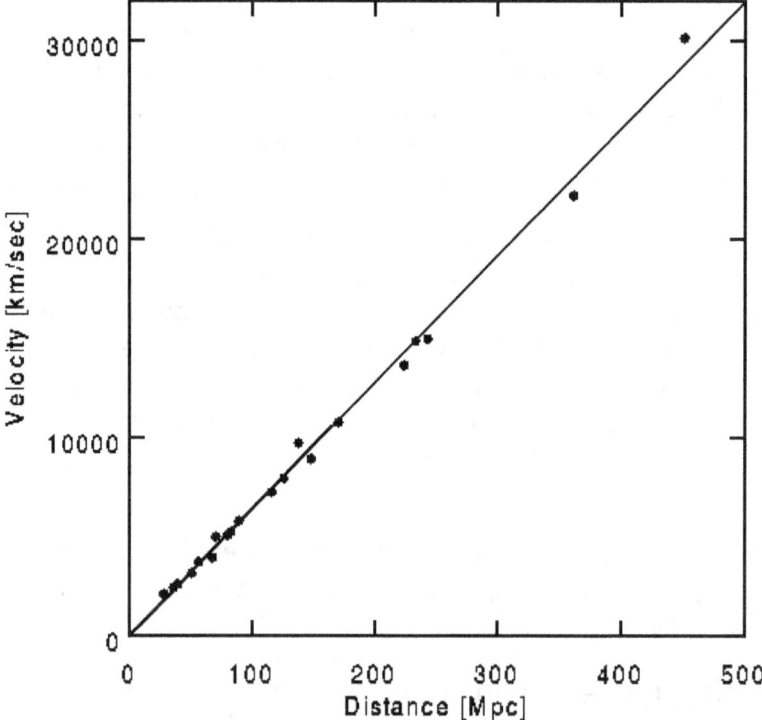

Figure 11-1 Hubble's Law
(data from Riess, Press and Kirshner, 1996)

The Law ... Used to Determine Distances

As explained in the last chapter, the Hubble law can be used to calculate distances of remote galaxies, once we have an accepted value for H_0 (we'll assume 74 km/s/Mpc). All we have to do is determine the velocity (v) of the galaxy and plug it into the equation, rearranged as follows:

$$d = \frac{v}{H_0}$$

And, as learned in Chapter 9, you determine the velocity by measuring the Doppler shift (Δw) and then applying the Doppler shift equation, rewritten below:

$$\frac{\Delta w}{w_0} = \frac{v}{c}$$

$$(\text{or}) \quad v = \frac{c\Delta w}{w_0}$$

where:

w_0 = wavelength when the source is not moving

Δw = amount of wavelength shift (in same units chosen for w_0)

v = radial velocity of moving source

c = speed of light (in same units chosen for v).

The Doppler shifts described in Chapter 9 applied to the fairly low stellar velocities within our galaxy. The shifts involved with remote galaxies are much larger and fundamentally different.

But I see Tri-face is getting a little restless, so I'll let him explain the fundamental difference between the normal Doppler shift and the shift we see for remote galaxies. OK, Tri-face, you're on center stage ...

Ahem ... ladies and gentlemen ... the Doppler shift that astronomers measure for individual stars in our galaxy (either blue-shifted or red-shifted) is caused by the star's motion *through* space. This is _not_ the same phenomenon (man, I love using that big word) as the redshift observed for remote galaxies. This redshift is caused by the *expansion of space itself* which sweeps the galaxies along with it and is therefore known as the **cosmological redshift**. Thank you for your attention. Any questions?

Sit down, Tri-face. There is no question period.

Here's an example of a cosmological redshift from a remote galaxy, NGC 3840, using the radio wave part of its EM spectrum. Galaxies contain regions of neutral hydrogen clouds in interstellar space that emit radio radiation at a wavelength of 21.12 cm. Because of its receding velocity, this radiation from galaxy NGC 3840 has been stretched (cosmologically red-shifted) to 21.64 cm in its EM spectrum. That's a wavelength shift ($\Delta w = w - w_0$) of 21.64 − 21.12 = 0.52 cm. So the galaxy's recessional velocity would be:

$$ v = \frac{c\Delta w}{w_0} = \frac{3 \times 10^5 \times 0.52}{21.12} = 7{,}386 \text{ km/s} $$

Imagine — a whole galaxy moving away from us at 7,386 kilometers every single second. That's roughly 100 times faster than the stellar velocities within our galaxy, but still only about 2.5% of the speed of light. The galaxy's distance from Earth can now be calculated as:

$$ d = \frac{v}{H_0} = \frac{7{,}386}{74} = 99.8 \text{ Mpc} $$

That's over 300 million light-years.

But when it comes to really remote galaxies, their recessional velocities are much more than 2.5% of the speed of light. For example, that supernova SN 1999fv, mentioned in the previous chapter — the one 9 billion light-years away — is receding from us at about 66% of the speed of light. And the example earlier in this chapter of a galaxy 13 billion light-years away is receding at 295,092 km/s which works out to over 98% of the speed of light.

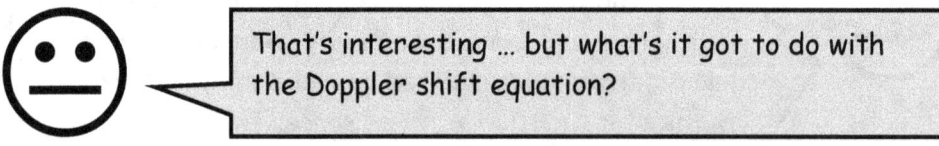

That's interesting ... but what's it got to do with the Doppler shift equation?

Well, determining the correct values for the enormous recessional velocities of really remote galaxies is not quite as straightforward as what we have shown so far. Whenever astronomers deal with velocities that are 10% of the speed of light or more, they need to pay attention to Albert (as in "Albert Einstein"). They need to allow for the effects of relativity to get the correct answer. Which means (sorry about this) they need a modified version of the Doppler shift equation.

Relativistically Speaking ...

But fear not! This is not nearly as scary as it sounds. And you'll actually get to use an equation that takes relativity into account. How cool is that?

But before we present the relativistic version of the Doppler shift equation, we need to introduce the z number shown on the left. It's just a convenient shorthand code that cosmologists like to use to represent the cosmological redshift that Tri-face talked about a couple of pages back. It's defined as the change (or shift) in wavelength (due to expansion of the universe) divided by the ordinary (unshifted) wavelength. What a mouthful. No wonder they find the z number a convenient code.

$$z = \frac{\Delta w}{w_0}$$

Using the z number, the usual Doppler shift equation shown here on the left, can now be written as the following simple linear equation shown on the right ...

$$\frac{\Delta w}{w_0} = \frac{v}{c}$$

$$z = \frac{v}{c}$$

... or, switching the two sides of this new equation to better compare it with the relativistic version coming up next, we have what you now see on the right. If you plot realistic values for v and z from this linear equation, you get a straight line.

$$\frac{v}{c} = z$$

Now, the relativistic version that astronomers use for this z relationship is as follows (please don't panic):

$$\frac{v}{c} = \frac{(z + 1)^2 - 1}{(z + 1)^2 + 1}$$

This version is not a linear equation, because if you plot realistic values for v and z from this equation, you get a curved line like Figure 11-2 where the value of z approaches infinity as the value of v approaches the speed of light (c).

NOTE: It's way beyond the scope of this book to explain how these geeky scientists derived this equation.

And besides, our illustrious author doesn't know.

Well, you got me there, Tri-face. Deriving relativistic equations is indeed beyond me. But that doesn't stop any of us ordinary folk from using some of the simpler ones. Like the one above. It may look a little complicated at first, but it's not. It's just simple arithmetic once you have a value for z. Let's have a go.

We're going to show you that relativity does make a difference — even when the recessional velocity involved is only about 2.5% of the speed of light. That was the velocity we calculated for the galaxy NGC 3840 on page 94 when we got an answer of 7,386 km/s. We're now going to recalculate that velocity using the relativistic version of the Doppler shift equation.

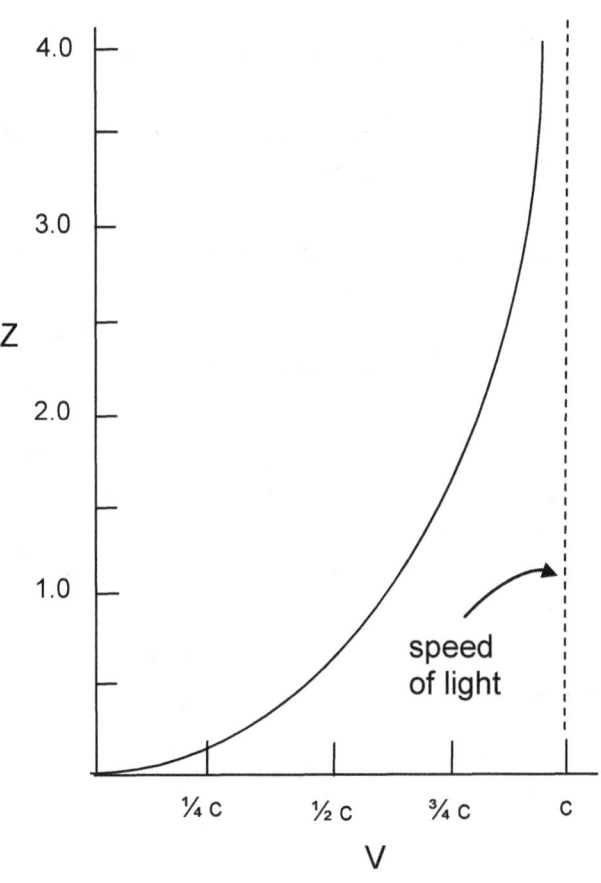

Figure 11-2 Relativistic Relationship Between z and v

For that galaxy (you may want to refer back):

$$z = \frac{\Delta w}{w_0} = \frac{0.52}{21.12} = 0.0246$$

So, plugging this z value into our relativistic equation on the next page, we get:

$$\frac{v}{c} = \frac{(z+1)^2 - 1}{(z+1)^2 + 1} = \frac{(0.0246+1)^2 - 1}{(0.0246+1)^2 + 1} = \frac{1.0246^2 - 1}{1.0246^2 + 1} = \frac{1.0498 - 1}{1.0498 + 1}$$

$$= \frac{0.0498}{2.0498} = 0.02430$$

Rearranging the result of the above equation by multiplying both sides by c gives us our revised value for the galaxy's recessional velocity:

$$v = 0.02430c = 0.02430 \times 3 \times 10^5 = 7{,}290 \text{ km/s}$$

This correct answer is 96 km/s lower than the more approximate value of 7,386 which did not take relativity into account. So ignoring relativity in this particular case has produced an error of:

$$\frac{96}{7290} \times 100 = 1.3\%$$

A fairly small error. But this sort of error gets bigger as these velocities get larger. Hence the guideline mentioned earlier that scientists should use the relativistic version of the equation when the velocities are 10% or more of the speed of light.

For a really far out distance and a huge recessional velocity, consider the supernova SN 1999fv mentioned in the previous chapter. It has a reported z number of 1.2, which is a pretty big cosmological redshift. If you care to plug this z value into our relativistic equation and do the arithmetic

$$\frac{v}{c} = \frac{(z+1)^2 - 1}{(z+1)^2 + 1}$$

... you will get a v/c ratio of 0.66. That means this object is moving at 66% of the speed of light (that's 198,000 km/s), so its distance according to the Hubble law would be:

$$d = \frac{v}{H_0} = \frac{198{,}000}{74} = 2{,}676 \text{ Mpc} \quad \text{(that's about 8.7 ly)}$$

The Law … Used to Determine the Age of the Universe

Try to visualize today's expansion of the universe from the time of the Big Bang by watching two remote galaxies that are widely separated from each other. Take their current separation as the distance d and the relative recessional velocity between them as the velocity v.

Now try to visualize the situation if time suddenly reversed. This would be like watching a film of the Big Bang expansion run backwards. The two galaxies would start moving toward each other at velocity v until they eventually covered the distance d and collided. That collision would represent the time of the Big Bang.

Here's the key point: The time taken to bring the two galaxies back together would represent the age of the universe from the moment of the Big Bang. This conclusion can be verified mathematically as follows (hang in with me here):

Let's call the time to bring the two galaxies together T_0. It's simply the distance between the two galaxies divided by the velocity at which they are coming together. In other words:

$$T_0 = \frac{d}{v}$$

To give a simple example of this equation — if the family car went a distance of 100 km at a speed of 50 km/h, it would take (100 divided by 50) hours to do so. That works out to two hours.

Now let's put some algebra to work for us. What happens if we use the Hubble law to replace the velocity in the above equation? Because the law is $v = H_0 d$, we would get:

$$T_0 = \frac{d}{H_0 d} = \frac{1}{H_0}$$

Well, well. Look what happened to d. It cancelled out. So the math is telling us that the value of T_0 does NOT depend on the distance d between the two galaxies. This means T_0 must be the same for all galaxies, no matter what their distance from each other. (That's because the greater the distance between any two galaxies, the greater the velocity between them, as indicated by the Hubble law.) So this T_0 must be the time when all the galaxies were crushed together — in other words, when the Big Bang took place.

This means that T_0 represents the age of the universe.

No wonder cosmologists have such a keen interest in the Hubble constant. The reciprocal of this constant (1 divided by H_0) represents the age of the universe.

I'm confused. You're telling me that the age of the universe is 1 divided by 74 kilometers per second per million parsecs? What kind of a time unit is that???

That's a fair question. The time unit is actually seconds, even though it doesn't look like it at first glance. If we rewrite this equation just using the units, we get:

$$T_0 = \frac{1}{H_0} = \frac{\text{no unit}}{\text{km/s/Mpc}} = \frac{s\,Mpc}{km}$$

I call this the flip-from-bottom-to-top rule. Because it's *per* s and *per* Mpc below the line (per is like division), we can move those two units to above the line. The example below, using arithmetic, illustrates that this crazy rule works. Here we use the slash(/) as a sign for per (or division). We start with:

$$\frac{50}{12/2/3}$$

We can do this arithmetic in one of two ways: (a) We can start by first doing all the divisions indicated below the line one step at a time. That's 12/2 to get 6, then 6/3 to get 2:

$$\frac{50}{12/2/3} = \frac{50}{6/3} = \frac{50}{2} = 25$$

... or (b) we can get the same answer using the flip-from-bottom-to-top rule by first moving the 2 and 3 to above the line (just like we moved s and Mpc to above the line in the T_0 equation):

$$\frac{50}{12/2/3} = \frac{50 \times 2 \times 3}{12} = \frac{300}{12} = 25$$

Now back to our T_0 equation. (Are you still with me?) We can plug in the 74 value for H_0 to calculate T_0, but we also have to convert the Mpc distance unit to km so that it cancels out with the km unit below the line. That will then leave us with just the unit s (seconds) for T_0. Now there are 3.09×10^{19} km in one Mpc, so the age of the universe is:

$$T_0 = \frac{1}{H_0} = \frac{1}{74 \text{ km/s/Mpc}} = \frac{1 \text{ sMpc} \times 3.09 \times 10^{19} \text{ km/Mpc}}{74 \text{ km}}$$

$$= 4.18 \times 10^{17} \text{ s}$$

Notice how the units Mpc and km cancel out, leaving only s.

OK, You've convinced me about the units. But isn't seconds a pretty stupid time unit to use for the age of the universe?

Well, that's the unit the equation gives us, Tri-face. But you should know by now that you are free to convert units to suit yourself. So if you would rather have the age of the universe expressed in years, just divide those 4.18×10^{17} seconds by the number of seconds in a year:

$$\frac{4.18 \times 10^{17} \text{ sec}}{365.25 \text{ days} \times 24 \text{ h} \times 60 \text{ min} \times 60 \text{ sec}} = 1.32 \times 10^{10} \text{ years}$$

That's 13.2 billion years, which is close to the currently accepted value of 13.7. As we mentioned in Chapter 6, analyze of the cosmic microwave background radiation that exists throughout space today provides a more reliable answer to this question than the Hubble law. The problem here is the nature of the Hubble constant.

The Constant That Isn't ...

The Hubble law approach for determining distances and cosmic age is too dependent on a realistic value for the Hubble constant. And besides, today's cosmologists don't consider it to be a "constant". They have good reason to believe that the universe's rate of expansion has changed over the course of its 13.7 billion history (more on this in Chapter 15). This means that each cosmic era of expansion would have its own, somewhat different, value of H_0.

Nevertheless, the Hubble law has been a valuable tool in our study and understanding of the cosmos ever since Edwin Hubble announced it almost a hundred years ago. And it continues to be so today.

Chapter 12 —
The Big Bang

It was Edwin Hubble's observations in the 1920s of an expanding universe that led to the Big Bang theory some years later. The theory deals with the beginning of the universe about 13.7 billion years ago and its early history. In this chapter, we will give only a very simplified description of this theory — a theory which has been formulated and refined over the last few decades by cosmologists and particle physicists. It's a theory that provides another striking example of why cosmology involves both the macro world and the micro world.

Understandably, a lot of non-scientists find the Big Bang theory hard to swallow. After all, nobody was around 13.7 years ago with a stop watch, thermometer, or digital camera. So there is no "proof", so to speak. It is really difficult to imagine the ridiculously short time frames and scorching temperatures immediately following the Big Bang that have been calculated and described by theorists. It's even harder to grasp how they can come up with such precise detail about such a unique event so long ago. But such is the power of science and math in the hands of competent people. This particular theory is just another example of the scientific method at work, as outlined in Chapter 8.

The Big Bang Theory

Let's be clear on two things about this theory. First of all, it appears to be a remarkably good theory because it explains a lot of things we observe today — for example, the expansion and structure of the cosmos and the large abundance of helium within it. The theory also predicts that there should be a **cosmic microwave background** radiation throughout today's universe at such and such a wavelength and such and such a temperature — because of (according to the theory) a transformation that occurred some 380,000 years after the Big Bang. Lo and behold, just such radiation was discovered in the mid 1960s, long after the prediction was made.

Secondly, this theory covers the early evolution of the universe from the time of the Big Bang. It says nothing about what caused the Big Bang. Some cosmologists argue that it is meaningless to ask what happened before the Big Bang because there was nothing prior to it. *Not even space and time.* Similarly, it is meaningless to ask what the universe is expanding *into* following the Big Bang because there is nothing beyond the universe. You may just as well ask what lies on the surface of planet Earth north of the north pole.

I feel a headache coming on.

Yeah, cosmology can do that to some people, Tri-face. You may not want to participate in chapter 16 where we ask what caused the Big Bang, then life.

From the Big Bang to the First Stars

At the very beginning of time (so the theory goes) all the matter and energy in today's universe was scrunched together to infinite density in what is called the **cosmic singularity**. (We'll explain singularities in Chapter 14 when we introduce black holes.) That's when some sort of cosmic "explosion" set everything into motion in a colossal expansion of space itself. The cosmos has been expanding ever since. Table 12-1 summarizes the main events that took place in that cosmic nursery right after the Big Bang.

As you will recall, the general theory of relativity treats space and time together as one continuous entity, called spacetime. However, in the micro world of quantum mechanics, the theorists tell us there are certain "quantum effects" that cause this smooth picture of space and time to break down for extremely small distances and extremely short time periods. In the case of time, this limiting value is 1.35×10^{-43} seconds. Consequently, our laws of physics break down at this point and do not allow us to know what goes on in shorter time periods than this.

So I have some bad news for you. No doubt you've been wondering (if not worrying) what the universe was like between time = 0 and time = 1.35×10^{-43} seconds. That's the first ten millionths of the first trillion trillion trillionth of a second in the life of the cosmos. It's probably kept you awake many a night. Unfortunately (sigh), general relativity cannot give us the answer. It seems nobody knows. Not the theoretical particle physicists. Not even Einstein.

But, as you can see from the table, once they get past that theoretical time limitation, theorists are quick to propose what probably happened from then on, starting with a scorching temperature of 10^{32} degrees C. That's over a trillion trillion times hotter than the core of the sun! At those temperatures, elementary particles were bumping into each other at close to the speed of light, so their kinetic energy was enormous. Hence the reference to the freight train in the table. Individual quarks could not stick together to form protons and neutrons for any length of time, because subsequent high energy collisions would immediately break them apart again. It was just too darn hot.

NOTE: **Kinetic energy** is energy of motion. The faster something moves, the more energy it has. That's why getting hit by a baseball traveling 100 km/h hurts a lot more than one traveling only 50 km/h. Same baseball, just a different speed. Speed makes a huge difference in the amount of kinetic energy (KE) because $KE = \frac{1}{2}mv^2$, where m is the mass of the object and v is its velocity. Double the velocity of an object and its kinetic energy becomes four times greater (because v is squared here). So when that velocity gets close to the speed of light, you can see why something as small as a subatomic particle would pack a lot of punch.

Time	Temperature (K)	Events
0 seconds	???	Cosmic explosion of infinitely dense matter and energy throughout all of space.
10^{-43} seconds	10^{32}	Elementary particles such as quarks and electrons collide with the energy of a 60 mph freight train.
10^{-6} seconds.	10^{13}	Free quarks combine to form protons (hydrogen nuclei) and neutrons.
3 minutes	10^9	Nucleosynthesis starts: Protons and neutrons combine to form nuclei of helium and a tiny amount of lithium and beryllium.
15 minutes	4×10^8	Nucleosynthesis ends.
380,000 years	3,000	H and He nuclei combine with electrons to form H and He atoms.
roughly 200 or 300 million years	a few 100	Clouds of H and He collapse under self gravity (as shown below) forming bigger and bigger clumps of gas which become hot enough to form the first stars.

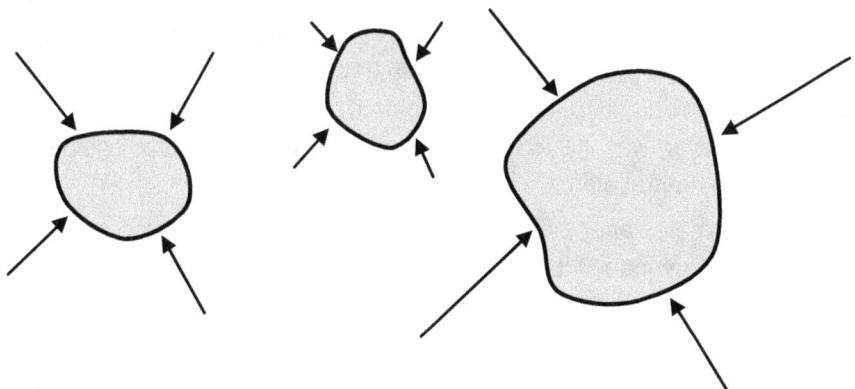

Table 12-1 A Hot Time in the Cosmic Nursery

About a millionth of a second after the Big Bang, the temperature had cooled down to about ten trillion K (10^{13} K). This reduced the energy of the high flying particles enough to allow quarks to stick together to form the protons and neutrons described in Chapter 2.

Roughly three minutes after the Big Bang, the temperature had cooled down to about one billion K (10^9 K), allowing nucleosynthesis to begin.

Allowing **WHAT** to begin???

Nucleosynthesis. It's the process of building up heavier atomic nuclei from protons and neutrons. It occurs when protons collide at high temperatures and pressures with other protons (and neutrons), causing them to fuse together into heavier nuclei. Because high temperatures are required, this process is also known as **thermonuclear fusion**. In this book we will refer to it simply as **nuclear fusion**.

Normally, two protons repel each other because they both have a positive electrical charge. But when the temperature is high enough, protons move so fast that their kinetic energy overcomes this electrical repulsion.

The reason nucleosynthesis didn't start earlier than three minutes after the Big Bang is that the temperature was actually too high at that time. Any formation of heavier nuclei was immediately broken up by other high flying energetic protons and neutrons.

A huge amount of helium (with two protons and two neutrons in its nucleus) was created during this Big Bang nucleosynthesis. This is why helium is by far the second most abundant element in the universe today (behind hydrogen). Only small amounts of lithium and beryllium formed before the temperature rapidly dropped. As you may recall from the periodic table in Table 2-1 on page 20, helium, lithium, and beryllium are the second, third, and fourth lightest chemical elements. The temperature, however, was still too high to allow individual electrons to join up with any nuclei to form neutral atoms.

Roughly 15 minutes after the Big Bang, the temperature had dropped sufficiently to shut down any further nucleosynthesis. The resulting matter consisted of positively charged nuclei (known as **ions**) and freely roaming negatively charged electrons. Such an ionized gas is called **plasma**.

For the next 380,000 years or so, energetic photons of electromagnetic (sorry, EM) radiation continually collided with the nuclei and electrons of the plasma. This further prevented electrons from binding with the nuclei to form hydrogen and helium atoms. The photons were absorbed by these collisions, so none of the EM radiation could get through the plasma.

Not until the temperature dropped to a mere 3,000 K about 380,000 years after the Big Bang were conditions right for the formation of atoms. But these newly formed atoms did not absorb the (now) lower energy photons, so these photons were free to move through space for the first time. Space became transparent to them. These same ancient photons are still hanging around today. They make up that cosmic microwave background radiation that was discovered in the mid 1960s.

During the millions of years that followed, the universe continued to expand and cool. Slight variations appeared in the amount of matter from one location to another within vast clouds of hydrogen and helium. This, in turn, created slight variations in the gravitational attraction in those locations — the more matter, the more gravitational pull. As a result, separate clumps of gaseous matter gradually formed. As the clumps got bigger, they attracted even more matter. More matter meant more gravitational pull, so the clumps continued to grow.

Eventually, a clump of matter would become so massive that the temperature and pressure in its central core — under all that weight of hydrogen-helium gas — would become high enough to start nucleosynthesis. It was just like the short nucleosynthesis period that started three minutes after the Big Bang.

As you will see in the next chapter, the tremendous amount of energy created by this nuclear fusion transformed the dense clump of gaseous material into a shining star. And so the universe suddenly lit up like a Christmas tree around 200 or 300 million years after the Big Bang. Stars appeared everywhere and eventually grouped together to form galaxies.

 But the nucleosynthesis from the Big Bang shut down after only 12 minutes. How come it didn't shut down in those newly formed stars? They just kept right on shining.

It shut down after the Big Bang because the young universe was expanding rapidly and therefore cooling down. The temperature quickly dropped below a level that could support the fusion process. In the newly formed stars, there was no such expansion and therefore no such drop in temperature. Internal gravity kept the mass of these stars compressed tightly together, thus maintaining hot temperatures and high pressures in their stellar cores.

Want to Dig Deeper?

This concludes our brief picture of the Big Bang. It leaves out some of the more complex and controversial issues within the theory. If you want to visit your local library or go online to dig deeper, you will encounter such issues as:

- the inflationary model which involves an extremely short lived but super fast period of cosmic expansion a tiny fraction of a second after the Big Bang
- the breaking out of the four fundamental forces from a single unified force
- the disappearance of antimatter.

 NOTE 1: The four fundamental forces of nature are (1) gravity, (2) the electromagnetic force, and two really short range forces within the atom: (3) the strong nuclear force and (4) the weak nuclear force. We have only discussed the first one (gravity) in this book.

NOTE 2: According to the theoretical particle physicists, an antimatter particle existed for every matter particle shortly after the Big Bang. These antiparticles had the same mass as their particle cousins, but the opposite sign for their other fundamental properties. For example, the electron's antiparticle was the **positron**, which was just like the electron except that it has a positive electric charge. Although some of these antiparticles have been produced by scientists in particle accelerators for fractions of a second, the universe seems to be void of any permanent antimatter today. This remains a bit of a mystery for scientists, because when a particle meets up with its antiparticle, they are both annihilated and all their mass is converted into energy. So if the Big Bang produced them in equal amounts, why do we have matter today, but no antimatter?

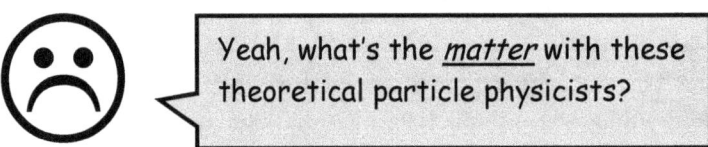

Yeah, what's the _matter_ with these theoretical particle physicists?

Ha ha.

Chapter 13 —
The Life and Times of Stars

At the end of the last chapter, we explained how stars are born via the gravitational contraction of matter within huge clouds of gas. We indicated that when the temperature gets high enough in the core of such clumps of matter, nuclear fusion is triggered. Over ten million degrees K is required. The result is a fusion of hydrogen into helium and the creation of a shining new star. At this stage, the gravitational contraction stops because it is balanced by the outward flow of the nuclear thermal energy from the core. The new star becomes physically stable and is said to be in **thermal equilibrium**.

But I've already said too much. Tri-face wants to introduce this chapter ...

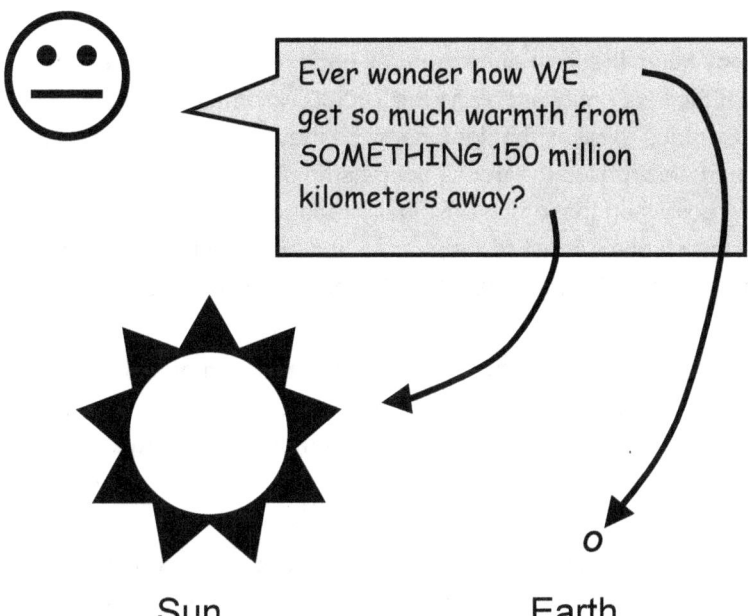

Sun   Earth

That's a really good introductory question, Tri-face. Ancient philosophers and scientists wondered about that for centuries. Next to the cosmos itself, the nature of our sun has always been a fascination for mankind.

For a while, various mythical sun gods were the answer. This was followed by some weird attempts at pseudo science that portrayed the sun as a wheel of fire, or a hot stone, or a giant mirror reflecting rays from a central fire at the heart of the universe. Then real science took over and energy from gravitational contraction was considered, and finally chemical combustion. But when astronomers first realized that our sun might be millions or even billions of years old, they realized that even if it was entirely composed of, say, carbon, it would have burnt through all its fuel a very long time ago.

What Keeps Our Sun Going?

We now know our sun is 4.5 billion years old and still going strong. So what could keep it going for so long, pumping out 3.90×10^{26} watts of power day in and day out? We know that nuclear fusion is going on in the solar core, but so what? How does that explain the incredible amount of energy produced?

Let's have a look. Through a series of reactions, the nuclear fusion process in the sun's core fuses four hydrogen nuclei into one helium nuclei ... _AND_ ... in so doing, also creates lots and lots of energy:

$$4 \text{ H} \rightarrow \text{He} + \text{ENERGY}$$

Most of this energy takes the form of gamma ray photons which take roughly 180,000 years to fight their way outward along the 700,000 kilometer journey from the core to the sun's surface. It takes that long because they bounce back and forth off zillions of nuclei and electrons in the dense hot plasma of the sun's interior, losing energy as they go. When these photons finally escape from the surface (presumably with a sigh of relief), most of them do so as visible light. They then deliver their warmth to Earth in 8.3 minutes traveling uninterrupted through space at the speed of light.

But the big question still remains. Why does this nuclear fusion of hydrogen into helium produce energy? And the answer lies with the laws of physics (otherwise known as mother nature) which demand conservation of mass and energy. The amount of mass and energy going into any reaction must equal the amount coming out.

When those theoretical geeks look at the masses involved in the conversion of hydrogen to helium, they find that the mass of the helium produced is actually 0.7% less than the mass of the original four hydrogen components. Where does this missing mass go? It gets converted to energy in accordance with the mass/energy conservation law.

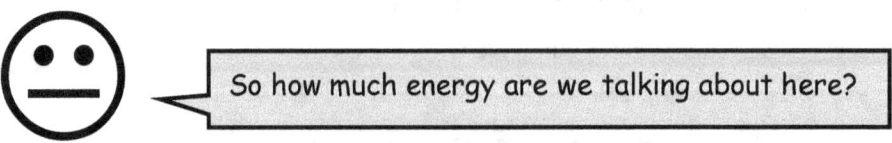

So how much energy are we talking about here?

This is where Albert Einstein comes into the picture (once again). Thanks to his theory of special relativity in 1905, we have what is no doubt the best known equation in all of science:

$$E = mc^2$$

where:

E = amount of energy (in J) into which m can be converted

m = amount of mass (in kg)

c = speed of light (3×10^8 m/s),

NOTE: J is the abbreviation for joules, a unit of energy mentioned in Chapter 9 when introducing luminosity and watts. We don't need to define joules quantitatively for this discussion.

One look at this remarkably simple equation and you will see that converting just a little bit of mass releases an awesome amount of energy. That's because c (a huge number to begin with) is squared in this relationship, becoming even bigger (9×10^{16}). The energy produced fusing hydrogen into helium is roughly ten million times more than what would be released in a typical chemical reaction, such as burning coal.

When Einstein published this equation over one hundred years ago, scientists realized that maybe it could — somehow — provide the answer to the sun's long life. Along with subsequent advances in theoretical nuclear physics, it didn't take these guys long to figure it out.

This stuff about huge amounts of energy from tiny bits of mass sounds awesome, but hard to get a feel for. How about an example?

OK. Let's say we want to convert one kilogram of hydrogen into helium. That's not very much hydrogen (about 2.2 pounds for those who still cling to that other system). According to the 0.7% loss of mass for this conversion, we only get back 0.993 kg of helium. The missing 0.007 kg of mass has been converted into energy as follows:

$$E = mc^2 = 0.007 \times (3 \times 10^8)^2 = 6.3 \times 10^{14} \text{ joules}$$

This amount of energy from only 0.007 kg of mass is equivalent to (are you ready for this?) burning 20,000 metric tons of coal! (One metric ton is 1,000 kg.)

Incredible. Now, if we get this much energy from converting just one measly kilogram of hydrogen to helium in the sun's core, how fast is the sun "burning" up this hydrogen nuclear fuel? All you need to understand here is that luminosity is a measurement of **power**, and power is the *rate of energy* being produced. Recall from our discussion of the inverse-square law on page 80 that one watt of luminosity is equal to one joule (J) of energy produced per second. So get your calculator ready …

---

1 kg of converted H in the sun's core produces 6.3×10^{14} J.
The sun's luminosity is 3.90×10^{26} W = 3.90×10^{26} J/s

Therefore … the rate of consumption of H in the sun's core is

$$\frac{3.90 \times 10^{26} \text{ J/s}}{6.3 \times 10^{14} \text{ J/kg of H}} = 6.2 \times 10^{11} \text{ kg/s (or) } 6.2 \times 10^{8} \text{ metric tons/s}$$

That's 620 million metric tons of hydrogen used up
EVERY SECOND !!!

Well, with that much "hydrogen burning" going on (as astronomers like to describe it), you might think that the sun's fuel would be used up in no time. But you'd be wrong. Check out the Reference Data section in the back of the book. The sun's total mass is about 2×10^{30} kg and most of this is hydrogen. Astronomers tell us that there is enough hydrogen in just its core to last another five billion years or so. It goes without say that without this energy from the sun, life on Earth would not be possible.

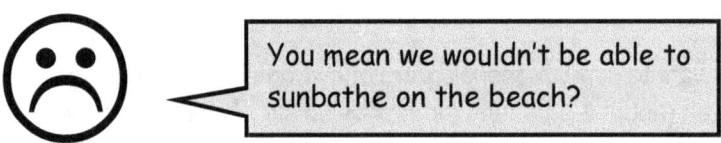

You mean we wouldn't be able to sunbathe on the beach?

No, I mean we wouldn't be here. Life needs energy and liquid water to survive. Without the sun, there would be no energy and all the water on the planet would be frozen solid.

Will Our Sun Ever Die Out?

Oh for sure. As mentioned above, in another five billion years or so, its core will run out of hydrogen and it will experience the stellar death described in the next chapter. And it will take Mercury, Venus, and Earth with it in a quiet, but fiery, ending. But life on Earth will have become pretty unbearable long before then because the sun is getting hotter all the time.

You might think that it would gradually get cooler and dimmer, like a camp fire running out of logs. But nuclear fusion in stellar cores doesn't work that way. As the hydrogen in the core gets converted into helium, the total number of atomic nuclei decreases (one helium replaces four hydrogen, right?). This causes the core to shrink, which in turn, makes it denser and hotter. And the denser and hotter the core, the faster the fusion process works. Which means an increase in luminosity. A hotter, brighter sun.

This trend will make life on Earth very uncomfortable, if not impossible, less than a billion years from now. It has been estimated that the sun currently has roughly 40 percent more luminosity than it had near the beginning of its life.

If the human race hopes to survive at that time, it will have to develop one super-duper giant air conditioning system. Or better yet, find a more comfortable solar system to move to.

Live Fast, Die Young

The more massive a star is, the more hydrogen fuel it has in its core. So you might think that it should live longer than a lower mass star, right? After all, if I have more gasoline in the tank of my car than you do in your family car, then — all else being equal — I will be able to drive farther than you will.

But in the stellar world, "all else" is not equal. The high mass star has a much higher temperature (and pressure) in its core than the low mass star. For example, our sun has a core temperature of 15.5 million degrees K, whereas a star 25 times more massive than our sun (a 25-M☉ star) has a core temperature of 40 million degrees K. This higher temperature causes nuclear fusion to proceed at a very much faster rate than in the low mass star. In fact, this rate is so fast that it uses up its core hydrogen much sooner than the low mass star, even though it had a lot more of it to start with.

In the stellar world, big guys live fast and die young.

Table 13-1 lists different sized stars (according to their mass), along with their surface temperature, luminosity, and lifetime expectancy. The masses and luminosities are expressed as solar masses and solar luminosities. The sun is included in the list (highlighted in gray) for comparison purposes. As you can see, mass has a profound effect on the characteristics of a star.

Standard Solar Model

The subject matter in this chapter is a theoretical picture of what goes on in the interior of stars. Scientists call it the **Standard Solar Model**. In Chapter 8 we discussed what scientists mean by the word "theory" ... how the acceptance of a

Mass (M☉)	Approximate Surface Temperature (K)	Luminosity (L☉)	Lifetime Before Running Out of Core Hydrogen (in millions of years)
25	35,000	80,000	5
15	30,000	10,000	15
3	11,000	60	500
1.5	7,000	5	3,000
1.0	6,000	1	10,000
0.75	5,000	0.5	15,000
0.5	4,000	0.03	200,000

Table 13-1 Characteristics of Stellar Masses

theory depends on the evidence that confirms it ... how a theory survives until proven false by some conflicting evidence.

Obviously, no scientist will ever visit the core of the sun to observe hydrogen nuclei fusing together. But rest easy. Astronomers have found overwhelming evidence in their various areas of research that supports the Standard Solar Model.

We'd like to close this chapter with just one example of such evidence. This example involves a fascinating story about missing neutrinos.

We mentioned neutrinos way back in Chapter 2. Just to remind you — they are elementary matter particles that come in three different types or flavors (as astronomers like to call them), all of which have no electrical charge and almost no mass. They move at almost the speed of light.

The Case of the Missing Neutrinos

When discussing the H-to-He reaction in the solar core, we left out a few minor details, one being the release of neutrinos.

The Standard Solar Model claims that for every helium nucleus produced, the reaction also releases two neutrinos. If scientists could somehow verify that this is true, it would be a major contribution to the supporting evidence for this theoretical model. So they decided to try. They calculated how many of these neutrinos should reach Earth each second according to the model. This flow of neutrinos at Earth's surface is referred to as **Earth neutrino flux** (EF). They then built several neutrino detectors around the world to measure this EF.

 NOTE: There was just one problem when it came to actually counting these little guys. Because they have no charge and hardly any mass, they are incredibly hard to detect. Most of them pass through matter without hitting anything (recall just how empty an atom is — see Figure 4.2 on page 30). They fly through the sun's interior unimpeded, reaching Earth 8.3 minutes later. Trillions of them pass right through our bodies (and indeed the entire planet) each second without any effect. As a result, the neutrino detectors presented some major engineering challenges. The detection material itself had to be extremely pure and designed to detect a certain (but incredibly small) percentage of the neutrinos passing through it. The detectors also had to be built deep underground to protect them from triggering false detections from things like radioactivity from nearby radioactive materials and cosmic radiation that constantly bombards Earth from outer space.

Let's look at the calculation. The sun's luminosity, L, is 3.90×10^{26} W (or joules per second). The Standard Solar Model indicates that the energy produced from one occurrence of fusing four hydrogen nuclei into one helium nuclei is 4.1652×10^{-12} joules. So we can calculate how many of these occurrences (N) take place in one second by dividing the first number (L) by the second number. Then the flow of neutrinos from the solar core would be 2N per second, since each occurrence of the reaction produces two neutrinos. This flow is called **solar neutrino flux** or SF.

$$SF = 2N = 2 \times \frac{3.90 \times 10^{26}}{4.1652 \times 10^{-12}} = 1.87 \times 10^{38} \text{ neutrinos per second}$$

This is such an enormous number of neutrinos leaving the solar core that a huge number of them (the EF) will still manage to pass through our tiny planet 150 million kilometers away. We can calculate EF for a given area of detector on Earth, recalling that the area of a sphere is $4\pi R^2$, where in this case, R = 1 AU or 1.50×10^{11} meters. This is just like those inverse-square law calculations in Chapter 10. In this case, SF is like luminosity and EF is like apparent brightness.

$$EF = \frac{SF}{4\pi R^2} = \frac{1.87 \times 10^{38}}{4 \times 3.14 \times (1.50 \times 10^{11})^2}$$
$$= 6.6 \times 10^{14} \text{ neutrinos/m}^2\text{/s}$$

So this is the EF that scientists hoped to find when they turned on their detectors. And that's when they started scratching their heads. The detectors recorded only a third to a half of the number of solar neutrinos predicted from the theory. A lot of solar neutrinos seemed to be missing.

Don't nobody leave this room!

Very funny, Tri-face. But the scientists were not amused. They decided there were only three possibilities for this unfortunate result:

1. The detector experiments were flawed.
2. The Standard Solar Model was wrong (perish the thought!)
3. Something strange happens to solar neutrinos after they leave the sun's core.

After checking and rechecking the detectors and the associated experimental procedures and reviewing the theoretical model once again, they concluded that the third possibility was the most likely. The big clue here is that there are three "flavors" (types) of neutrinos — electron-neutrinos, muon-neutrinos, and tau-neutrinos.

The theory specified that the neutrinos produced in the solar core were electron-neutrinos and the detectors had all been designed to detect that type. Besides, the other two types were more difficult to detect. But some theorists had reason to suspect that these electron-neutrinos might transform into the muon or tau types when passing through matter — such as the solar interior and our planet at night (that is, when the detector is on the opposite side of the globe from the sun). This transformation effect (as yet unproven) was called **neutrino oscillation**.

Around the turn of the century, a new detector was build with a different technique designed to investigate neutrino oscillation. This technique enabled the detection of all three types. The results were quite exciting. The data showed excellent agreement with the theoretical EF figure. It turned out that more than half of the original electron-neutrinos from the solar core went through oscillations, arriving at the detector in the form of muon or tau neutrinos. The case of the missing neutrinos was solved.

Aftermath of the Missing Neutrino Solution

Science marches on. The consequences of this solution were twofold:

1. The Standard Solar Model received some additional support by providing experimental confirmation on the release of two electron-neutrinos in the hydrogen-to-helium fusion reaction.

2. The Standard Model (mentioned in Chapter 2) that deals with particles in the micro world had to be changed to reflect the existence of neutrino oscillation. Prior to this, the theory had proposed that neutrinos had no mass. But it seems that some small amount of mass is theoretically required to enable the occurrence of neutrino oscillation. Scientists are now trying to determine the actual values of those masses for all three flavors of neutrinos.

Chapter 14 —
The Death of Stars

Strange as this may sound, it is the death of stars over the past millions and billions of years that has made life possible on our planet earth (and who knows where else). This is just one reason why scientists are interested in the morbid subject of dying stars.

Think back to the periodic table of chemical elements, a small portion of which appears in Table 2-1 on page 20. There are 92 known natural elements in the universe, but so far in this book we seem to be spending most of our time talking about just two of them — hydrogen and helium.

That's not my fault.

Of course not. Nobody said it was.

As everybody knows — hydrogen and helium are the two lightest and most plentiful elements in the universe. They were essentially the only two produced a few minutes after the Big Bang, before cooler temperatures prevented any more nuclear fusion. And as we have just learned in the last chapter, the next occurrence of nuclear fusion in the history of the universe was the H-to-He fusion taking place in newly formed stars with core temperatures of ten million degrees K or more.

So the question is for us humans — fragile creatures that we are, tucked safely away on a cozy planet in a quiet corner of a very hostile universe:

How did we get here in the first place?

Or, more to the point:

Where did the atoms in our bodies come from?

Our bodies contain lots of atoms besides hydrogen — atoms such as carbon, oxygen, nitrogen, sodium, magnesium, chlorine, potassium, and calcium. Even iron. The surprising answer to the above question is that all these atoms originally came from dying stars. That's the quick answer. Read on to get the whole story ...

Low Mass Dying Stars

NOTE: **Low mass** stars are loosely defined as having an initial mass at birth of roughly 8 M☉ or less. That includes our sun, of course, since it has a mass of 1 M☉. Stars more massive than 8 M☉ are known as (you guessed it) **high mass** stars.

When a low mass star runs out of hydrogen fuel in its core, it marks the beginning of the end. Thermal equilibrium is temporarily suspended as gravity takes over and the core starts to collapse. But compressing the core raises its temperature. When it reaches around 100 million degrees K, the temperature is high enough for the helium "ash" from the "hydrogen burning" to become nuclear fuel for the fusion of helium ("helium burning") into mostly carbon and oxygen. This restores thermal equilibrium until the core runs out of helium, resulting in a further collapse of the core.

Well, what da ya know. Carbon and oxygen! Finally — the creation of chemical elements heavier than helium. But why does the core temperature have to be so much higher before helium nuclei will fuse together to form these heavier nuclei?

Recall, Tri-face, that very high temperatures are needed in order to move the hydrogen nuclei (single protons) fast enough for them to overcome the repulsive electric force between them and fuse together. But this repulsive force is twice as strong when trying to fuse helium nuclei together, because each helium nuclei has two protons in it. Therefore higher temperatures are needed. As it turns out — much higher.

The sequence of events in a dying star is actually more complicated than described here. But basically, the scenario of core collapse followed by renewed fusion (helium burning) results in a gradual expansion and cooling of the star's atmosphere. The expansion is quite large, sometimes pulsating back and forth, but the end result is known as a **red giant**. Red giants have diameters from ten to a thousand times that of the sun, but also a lower surface temperature, namely 2,000 to 4,000 K. This lower temperature gives these dying stars their red appearance. Eventually, their atmosphere drifts away from the collapsed core into interstellar space.

NOTE: This enormous stellar expansion is why the sun will swallow up Mercury and Venus and burn Earth to a cinder when it becomes a red giant.

Low mass stars bigger than roughly 4 M⊙ have core temperatures of about 600 million degrees K after helium burning and subsequent core collapse. That's high enough to push the nuclear fusion process even further. "Carbon burning" fuses carbon nuclei to produce more oxygen as well as elements such as neon, sodium, and magnesium.

When low mass stars can no longer continue these higher fusion processes in their red giant stage, they are left with only a solid, hot shrinking core. This core contains very densely packed nuclei (mostly carbon and oxygen) and unattached ("free") electrons. The star's outer layers get ejected into space as an expanding cloud of gas known as a **planetary nebula**. This is a strange choice of name because the nebula has nothing to do with planets. But — as everybody should know by now — astronomers are strange folks. Eventually this nebula fades into the interstellar media of surrounding gas and dust.

The hot compressed core is known as a **white dwarf** because it glows white hot before cooling down over millions of years. It's only about the size of planet Earth and its mass is something less than 1.4 M⊙. Free electrons in the white dwarf are compressed to the point where they resist any further shrinking. The resulting matter is much more compact than you would have with a normal atomic structure. As Table 4-1 tells us on page 33, one coffee mug of this stuff would weight 500 tons here on Earth.

Think of a white dwarf as the corpse of a low mass star that died relatively quietly. Tri-face and I are reserving a cemetery tomb stone (Figure 14-1) for the day when our very own sun sadly passes away. On this stone, B reveals the birth and death dates for our beloved sun in Billions of years after the Big Bang.

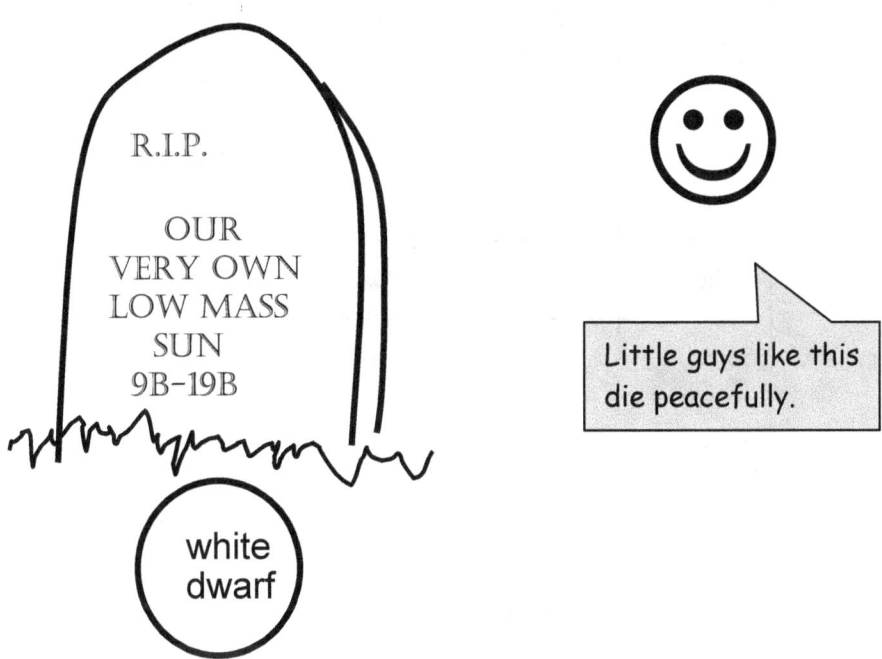

R.I.P.

OUR
VERY OWN
LOW MASS
SUN
9B–19B

white dwarf

Little guys like this die peacefully.

Figure 14-1 The Death of Our Sun

High Mass Dying Stars

The story is quite different for high mass stars because they have much higher core temperatures. Like the Standard Solar Model for living stars, what follows here is based on theoretical models. However, these models seem to be in remarkably good agreement with astronomers' observations of supernovae — the subject we are leading up to here.

These big guys do not go quietly to their grave. This turns out to be an explosive (ha ha) story.

Because of these higher temperatures, these stars fuse heavier nuclei together than in low mass stars during the dying process. The reactions become quite complex and a great deal of the released energy is in the form of neutrinos streaming out of the core. Eventually, silicon is fused into a variety of nuclei right up to iron at which point there is no further nuclear fusion. The iron nucleus is so stable that it can't be used as fuel in an *energy-releasing* fusion reaction. Elements heavier than iron can only be produced by *inputting* energy into the reaction ... as we shall soon see ...

	Burning Stage	Temperature of Core (K)	Density of Core (kg/m^3)	Duration of Stage
living star	hydrogen	4×10^7	5×10^3	5×10^6 years
dying star	helium	2×10^8	7×10^5	7×10^5 years
	carbon	6×10^8	2×10^8	600 years
	neon	1.2×10^9	4×10^9	1 year
	oxygen	1.5×10^9	1×10^{10}	6 months
	silicon	2.7×10^9	3×10^{10}	1 day

Table 14-1 Stages in the Life of a 25-M⊙ Star

Each new stage of nuclear fusion shrinks the core further and burns the "ash" products from the previous stage at a much higher temperature. So the burning goes faster from one stage to the next and the core continues to compress. By the time a high mass star reaches the final stage or two, these reaction occur extremely fast, using up the associated nuclear fuel within days. In contrast, the first stage or two takes many thousands of years. Table 14-1 shows some amazing data in the evolution of a star 25 times more massive than our sun. Have a close look at the figures.

This table presents a startling example of the "live fast, die young" principle mentioned in the last chapter. While our sun will have a life span of ten *billion* years, this monstrous 25-M⊙ star has a life span of less than six *million* years.

As mentioned above, to create elements heavier than iron, nuclear reactions require an input of energy. This type of reaction is quite a different game, but one that high mass stars play as soon as the nuclear fusion in their core ends with iron. The game becomes their final, spectacular act on the cosmic stage.

Look Out, Here Comes a Supernova!

The moment the silicon burning is completed in the iron rich core of a high mass star, thermal equilibrium comes to an abrupt and final halt. Internal gravity takes over in earnest. In the next quarter of a second (according to theory) the dense core collapses even further, shrinking to a diameter of roughly 25 kilometers — the size of a city. This is much smaller than a white dwarf because, unlike the white dwarf, it's mass is greater than 1.4 M⊙. When gravity tries to compress a mass greater than 1.4 M⊙ even further, it succeeds in doing so because it is strong enough to break down the resistance of the free electrons.

During this core collapse, the negatively charged electrons are forced to combine with the positively charged protons to form neutrons. The core is now at **nuclear density**. This is the density that you find in the nucleus of atoms, where protons and neutrons are packed tightly together. This explains the 4×10^{17} kg/m^3 density given back in Table 4-1 on page 33, where a coffee mug of this stuff would weight 50 billion tons here on Earth. There is no place in this core for the ridiculously roomy atomic structure we saw in Figure 4-2 on page 30. Only the resistance of the free neutrons prevents further shrinking below 25 kilometers.

Now the show begins. As gravity tries to compress the nuclear dense core even more, this core suddenly becomes very rigid, sending an enormously powerful shock wave of high pressure *outward*.

Meanwhile ... back in the region outside the core, gravity is drawing the material there *inward* at tremendous speeds. When this inward material crashes into the outward shock wave, it is forced back again toward the surface of the star. The wave accelerates on its upward journey as it gradually meets less residence, reaching the star's surface in a matter of hours. The outer layers of the star eject into space. The energy created from the collision between the outbound wave and inbound stellar material has been estimated at a colossal 10^{46} joules — a number that staggers the imagination. The star, in effect, has blown apart in a gigantic explosion.

The amount of energy quoted above is about a hundred times more than all the energy emitted by our sun since its birth 4.5 billions ago. You may want to read that sentence again.

That's some firecracker!

Well, yes, you could say that. But it's really a **supernova**. The awesome amount of released energy escapes from the accident scene in the form of neutrinos and light. For a month or two the luminosity from a core-collapsing supernova can be as high as 10^9 L⊙ which — as everybody now knows — means a billion times brighter than our sun. The light is so intense that it shows up as an individual "star", even in a galaxy that is very remote. This light can last for many days before slowing fading away.

NOTE: Astronomers identify different types of core-collapsing supernovae, according to the presence of hydrogen and helium lines in their spectra. However, the Type Ia supernova that cosmologists use as a distance indicator (see Chapter 10) is quite different. Rather than core collapse, Type Ia gets its energy from a runaway thermonuclear explosion on a white dwarf. This happens when the white dwarf is a partner in a binary star system and takes on too much material from its nearby companion star, thus exceeding its 1.4 M⊙ limit. As a standard candle for determining distances, Type Ia has two big advantages over core-collapsing supernovae — its luminosity is greater and it's also constant in value for all occurrences of this type of supernova.

Bring On The Heavy Metal

Great! I could use some hard rock music about now.

Nice try, Tri-face, but it's not that kind of heavy metal. You may recall that astronomers refer to all the chemical elements that are heavier than helium as "metals". If we now consider all those "metals" up to iron (atomic number 26) in the period table as *light* metals, then we are about to learn where the *heavy* metals in the universe come from — those from cobalt (number 27) all the way up to uranium (number 92).

The environment created by the supernova shock wave contains such an unbelievable concentration of energy that the stellar material within it is subjected to a whole new series of thermonuclear reactions that use this energy as input. The result is the forging of all the heavy metals in the period table. As it turns out:

There is no other environment in the universe where these heavy metals can be produced. This is another sentence you may want to read again.

We now have a more detailed answer to the question posed at the beginning of this chapter: *Where did the atoms in our bodies come from?* They came from quietly dying low mass stars and exploding high mass stars.

The lighter elements came from all kinds of dying stars, big and small (including supernova size), but the heavier ones came only from supernovae. So just imagine — some of the atoms in your body came from supernova explosions that took place before our solar system was formed 4.5 billion years ago. In fact, the atoms in your body probably came from several different generations of stars that lived billions of years ago. *If there were no supernovae, we folks would not be here.* Wrap your mind around that!

So as dramatic as this may sound, it is absolutely true when astronomers say:

We are all made of star dust.

Well, this is all quite amazing, but how did all those elements actually get from these stars (that existed heaven knows how far away or long ago) to our little planet where life evolved?

Good question. A low mass star the size of our sun loses, very roughly, 40 percent of its material to interstellar space during its death throws. Some of that material includes light "metals" like carbon and oxygen, because there is a lot of churning (or **convection**) of the star's internal material at that time. This churning dredges up some of the nuclear products from the core and distributes them all the way to the star's surface.

Computer models of supernova explosions indicate that a 25 M⊙ star like the one in Table 14-1 loses up to about 96 percent of its material to interstellar space. This material includes a variety of light metals and all the heavy metals in the periodic table.

All this lost material becomes available in interstellar space as gas and dust for the formation of the next generation of stars. These new stars, therefore, contain not just hydrogen and helium at their birth, but also "metals".

The amount of metals in a star is a measure of the stars **metallicity**. As the universes ages, this natural process of stellar birth and death causes each generation of new stars to have higher metallicity than the previous one. Since planets are formed from the same material that formed their sun, a metal-rich star will have metals in the planets closest to it (if not in the ones farther out). Our sun is a metal-rich star, which is why we find heavy elements on our planet right up to uranium.

Neutron Stars

It's only fitting that we conclude this chapter on dying stars by talking a little more about corpses. Not those dense little white dwarfs from low mass stars, but those extraordinary dense, tiny ones from high mass stars.

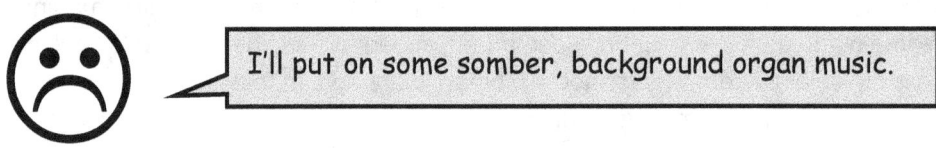

I'll put on some somber, background organ music.

Sure. Whatever.

The dense core that a supernova leaves behind as a corpse is known as a **neutron star**, because it is composed mostly of neutrons. This unique stellar object at roughly 25 kilometers diameter can have a mass between 1.4 M⊙ up to about 3.0 M⊙. That's like three of our suns all scrunched together into the size of a city. These so-called "stars" are pretty weird.

Because of their extreme density and small size, they can rotate up to 1,000 times per second and have enormous magnetic fields. Astrophysicists have developed theoretical models of neutron stars in an effort to understand their structure. The results are rather uncertain to say the least, but indicate, among other things, a very thin solid surface crust of heavy nuclei and freely roaming electrons and an interior of mostly superconducting, superfluid neutrons that flow without causing any friction.

You would not want to hang around on the surface of this thing. But you wouldn't have much choice if you magically found yourself there, because its escape velocity is enormous. In Chapter 1 we mentioned that each celestial object has its own escape velocity. The more massive the object and smaller the size, the denser it is — and therefore the stronger the gravitational field at its surface. Which means a higher escape velocity. We could simply tell you what this velocity is for a neutron star, but it might be more interesting to actually calculate it. Just another example of how algebra comes to the rescue.

Physicists have derived the following handy dandy equation for a celestial object:

$$v_e = \text{square root of } \frac{2Gm}{r}$$

where:

v_e = escape velocity (in km/s)

G = universal constant of gravity $(6.67 \times 10^{-11} \text{ N m}^2/\text{kg}^2)$

m = mass of the object (in kg)

r = radius of the object (in m)

 NOTE: You saw G in Newton's universal law of gravitation in Chapter 1. We need its value here, but don't worry about the units. (If you must know, N stands for newton, a unit of force.)

We reported back in Chapter 1 that the escape velocity was 11.2 km/s for Earth and 617.4 km/s for the sun. Compare this with what follows here. Since 1 M⊙ is 1.989×10^{30} kilograms (see the Reference Data section in the back of the book), a 25 kilometer diameter neutron star with a mass of 2 M⊙ would have an escape velocity of:

$$v_e = \text{s.r. of } \frac{2Gm}{r} = \text{s.r. of } \frac{(2 \times 6.67 \times 10^{-11}) \times (2 \times 1.989 \times 10^{30})}{12.5 \times 10^3}$$

$$= 2.06 \times 10^8 \text{ m/s (or) roughly 200,000 km/s}$$

The speed of light is 300,000 km/s, so to escape from a neutron star like this you would have to leap off at about two thirds the speed of light. Forget it.

But if you think this is weird, guess what the theorists tell us about a high mass star so big that the resulting neutron core exceeds 3.0 M⊙.

Black Holes

If the core is more massive than 3.0 M⊙, its internal gravity overcomes the resistance of the free neutrons and the core collapses down to a single point of infinite density. You may recall in Chapter 12 that the Big Bang originated from a point of infinite density.

 Infinite density? You're pulling my leg, right? Density is mass divided by volume. My math teacher tells me when you divide something by zero you get infinity. So you're telling me the volume of this "singularity" must be ... like ... zero. Which means it has no dimensions. So this mass takes up no space? HELLO!!!

I told you this was going to be weird. But that's what a "point" is — just a location with no dimensions. In this particular case, the point happens to have mass. A lot of mass. Astronomers call it a **singularity**. It represents the center of a **black hole**. A certain distance out from the singularity is an imaginary spherical boundary known as the **event horizon**. It represents the outer boundary of the black hole. Anything that ventures too close to this horizon gets pulled across it by the enormous gravitational field there and swallowed up by the singularity. Just such a disaster is indicated by the heavy arrow for that unfortunate lost soul in Figure 14-2.

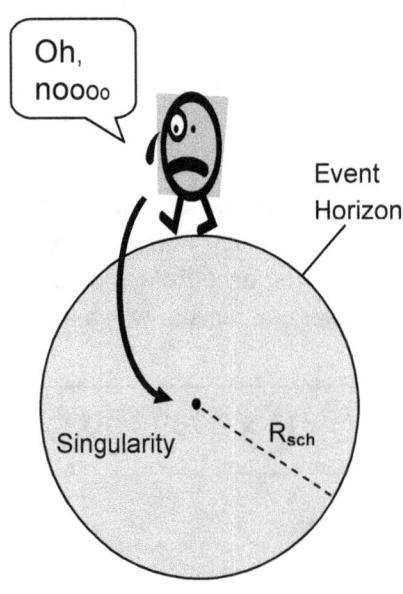

Figure 14-2 A Black Hole

The theoretical guys tell us that our laws of physics break down inside a black hole because the gravitational field at the singularly is infinite. This means an infinite curvature of spacetime. Space and time get all mixed up once you move inside the event horizon. It's as if the collapsed core from the massive star disappeared from the universe. In Chapter 8 we talked about a depression in spacetime called a **gravity well** that surrounds massive objects. For a black hole, this gravity well is infinity deep.

How would you like to figure out the size of a black hole? You could tell your buddies all about it.

There is a simple formula that allows us to calculate the radius of a black hole — the distance from the central singularity to the event horizon. It was derived by the German astronomer, Karl Schwarzschild who must have been a really bright cookie. He became the first mortal to solve Einstein's equations of general relativity back in 1916. The radius has been named the Schwarzschild radius. (I would have preferred the Karl radius, but nobody asked me.):

$$R_{sch} = \frac{2Gm}{c^2}$$

where:

R_{sch} = Schwarzschild radius of black hole (in m)

G = universal constant of gravity (6.67×10^{-11} N m^2/kg^2)

m = mass of black hole (in kg)

c = speed of light (3×10^8 m/s).

Let's apply this equation to the black hole of a collapsed core of mass 4 M⊙

$$R_{sch} = \frac{2Gm}{c^2} = \frac{(2 \times 6.67 \times 10^{-11}) \times (4 \times 1.989 \times 10^{30})}{(3 \times 10^8)^2}$$

$$= 11{,}793 \text{ m} \quad (\text{or}) \quad 11.8 \text{ km}$$

That's about the size of the core before it collapsed into a black hole. Interesting.

Anything that goes into the black hole is trapped there. It can never return because the escape velocity at the singularity is infinite (in this case, r = 0 in the equation for escape velocity on page 125). This means that not even light can escape from the black hole. It's trapped within the event horizon. So we can never actually see a black hole. That's why it's called "black".

The existence of black holes was predicted by Einstein in his general theory of relativity. But even though black holes emit no EM radiation of any kind, astronomers now have ample evidence that such things really do exist. Their strong gravitational presence causes visible celestial objects to swirl around them at great speeds. By observing this sort of dynamic activity, astronomers have been able to calculate the mass and location of the unseen object that is driving this frantic motion.

Black Holes Come in Different Sizes

There is now considerable evidence for the existence of two quite different sizes of black holes. There is the stellar-mass size we have been discussing, which amounts to a small handful of solar masses, depending on the size of the original high mass star. These guys have been identified as members of binary star systems in which their smaller companion star has been observed orbiting around "an unseen object" at enormous speed. Calculations indicate that such speeds could only be caused by the gravitational influence of a black hole.

Then there are the "supermassive black holes" that lie in the center of many (if not all) large galaxies. Presumably the dense population of stars that gather in the center of a galaxy eventually leads to a progressive merging of such stars. When the merger reaches a certain mass and its gravity compresses that mass into a small enough volume, it gives in to the laws of general relativity by becoming a black hole.

It is estimated by the speeds and locations of stars and gas swirling around the center of our Milky Way galaxy that it contains a supermassive black hole of about 2.6 million solar masses. Material swirling around other galactic centers and emitting high levels of radiation indicate that supermassive black holes are probably quite common and range from a million to a billion or more solar masses.

Surprisingly, supermassive black holes are not that difficult to form. Theory tells us that the more mass involved, the lower becomes the required density to cause collapse into a black hole. So these supermassive black holes spread out over quite a large volume at quite a low density. The Schwarzschild radius for a billion solar mass (10^9 M☉) black hole would be:

$$R_{sch} = \frac{2Gm}{c^2} = \frac{(2 \times 6.67 \times 10^{-11}) \times (10^9 \times 1.989 \times 10^{30})}{(3 \times 10^8)^2}$$

$$= 2.95 \times 10^{12} \text{ m} = \text{(or) } 2.95 \times 10^9 \text{ km (or) } 19.7 \text{ AU}$$

Our old friend and ex-planet Pluto, now sulking out in the boon docks beyond Neptune, orbits the sun at an average distance of 39.5 AU, so this gigantic black hole is about half the size of our solar system.

Black Holes Don't Suck

On a final note to close out Part II of this book — some people worry about getting sucked in by a black hole. I know Tri-face does. They worry that maybe some day that mass-hungry monster in the center of our own galaxy might grow out of control and do us all in. Fear not. A black hole is just another form of mass. You have to get really close to one (like that idiot in Figure 14-2) before its gravitational field will have any noticeable effect on you.

Think of it this way. If you could magically turn the sun into a black hole, it would still have the same mass, just a much smaller size. So the motions of the planets would be virtually unaffected. The solar system would happily carry on as usual. What size would the sun be as a black hole, you ask?

Well, we've given you the equation that provides the answer. Work it out. If you don't get something around three kilometers, I'll be really disappointed. Especially after doing all this coaching.

Part III

Unsolved Cosmic Mysteries

Chapter 15 —
What's All This Dark Stuff?

In Part III we will briefly describe some of the cosmic mysteries that scientists have yet to solve. If you are thinking of pursuing science and math academically in the coming years, perhaps you too will end up working on some of these unanswered questions. Who knows, you might be the next Stephen Hawking. Scientific research can be a fascinating and rewarding challenge for anyone who has an interest and aptitude in that direction.

At the philosophical level, however, you may conclude that some of these mysteries lie beyond the reach of science. And you could be right. After all, it's hard to imagine how we could ever observe, say, another universe somewhere beyond our own. Perhaps lurking in some hidden dimension. While the men and women in the white lab coats can speculate on possible answers for some of the tougher questions, they may never be able to verify those answers with observations or experiments. That kind of speculation would never attain the status of a scientific theory.

Anyway, have a look at these cosmic mysteries with an open mind. And remember — if in doubt about a future in science — you can always take up philosophy instead.

Dark Matter

It's called **dark matter** because it has no luminosity. It sheds no light. So nobody can actually see this stuff. They can only observe its gravitational effect on the visible matter in its vicinity. That's how they know *something* is there.

What We Know

This mystery has been gradually surfacing since the 1930s and is based on measurements of celestial velocities via the Doppler effect (described in Chapter 9). Using this technique, astronomers have determined the velocity at which galaxies rotate at different distances from their center. They have also determined the velocity at which individual galaxies move within clusters of galaxies. In both cases, these velocities are much higher than the theoretically calculated values that are based on the gravitational effect of the visible matter in these galaxies and clusters. We're not talking relativistic speeds here, so Einstein's general theory of relativity is not needed. These calculations are based on Newton's universal law of gravitation.

To account for these observed velocities, astronomers figure there must be roughly ten times the amount of matter in and around these galaxies than is visibly present— extra matter that seems to exert the normal pull of gravity but has no detectable luminosity.

What We Don't Know

Nobody knows what this dark matter is, even though calculations suggest that it makes up about 90 percent of the matter in the universe.

That's gotta be embarrassing for cosmologists when some smart aleck asks them about it at a party.

What Might Be

Cosmologists have a few ideas. Some of the dark matter might be ordinary matter that is just too dim or otherwise unobservable. Such matter would include celestial objects beyond our solar system like planets, moons, black holes, neutron stars, white dwarfs, brown dwarfs (star-like objects not massive enough to become hydrogen-burning stars), and zillions of those tiny neutrinos. However, estimates indicate that all this "dim" ordinary matter could account for only a small portion of the so-called dark matter.

Other speculations call on the micro world of particle physics. They include (1) lightweight particles moving at high speeds, dubbed **hot dark matter** (one example, known to exist, being neutrinos) and (2) a whole new set of exotic particles, not known to exist, that are massive and slow moving, dubbed **cold dark matter**. It is thought that these latter particles don't interact much with ordinary matter (except for their mutual gravitational attraction). So they are known as <u>w</u>eakly <u>i</u>nteracting <u>m</u>assive <u>p</u>articles or **WIMPS** for short (I kid you not). So far, however, there is no sign of any WIMPS in any particle accelerator experiments.

Because this stuff has no detectable luminosity, it's probably quite different in nature from ordinary matter which is made up of good old protons, neutrons and electrons.

WOW! I could be way out to lunch on this, but this suggests to me that a radical modification may be needed to the Standard Model of particle physics.

Cool, Tri-face. You really are paying attention. And you could turn out to be right, if they ever get to find any of these elusive particles in space or in their particle accelerator experiments. As pointed out in Chapter 8, you never know when a theory may have to be altered or replaced. The cosmos is stranger than we can imagine.

And here's one final piece of speculation. An hypothesis was proposed in 1983 called Modified Newtonian Dynamics (MOND) that suggests there is no dark matter.

There is, instead, a flaw in Newton's universal law of gravity when applied to the conditions within a rotating galaxy. The MOND model offers a modification to Newton's law to account for the observed rotational velocities. MOND, however, leaves some basic issues unanswered and has not received much support in the scientific community.

Dark Energy

It's probably called **dark energy** just to put it in the same category as dark matter — a mysterious invisible unknown. Cosmologists can only observe its "gravitational" effect on the expanding universe. That's how they know *something* is there. But in this case, its actually an antigravity effect.

What We Know

This mystery surfaced in 1998. Up until then, cosmologists had always assumed that the rate of expansion of the universe (as expressed by the Hubble law) is gradually slowing down.

Because the enormous outward force from the Big Bang, causing the mass of the universe to fly apart, is challenged by the inward force of gravity. This inward force is trying to pull all that mass back together again. After all, that's what gravity does for a living, right? It pulls stuff together. And by doing so, this should naturally cause the outward expansion to slow down. A very reasonable assumption.

 NOTE: We need a brief aside here to explain why the rate of cosmic expansion — and any change to that rate — is so important to cosmologists. If there is enough mass in the universe to create enough gravity to eventually bring the slowing expansion to a halt, then gravity will win this tug of war and start pulling everything back together again. Eventually, everything will merge together in a Big Crunch, probably ending up as a cosmic singularity. Sound familiar? A cosmic singularity was the starting point of the original Big Bang, as outlined in Chapter 12. On the other hand, if there is not enough mass in the universe to bring expansion to a halt, then the universe will expand forever into an ever colder, darker, emptier place. Cosmologists care about such things, even if they can't do anything about it. That's just the way they are.

So in 1998, two teams of cosmologists, working separately, investigated this cosmic expansion rate. They perfected the technique of using Type Ia supernovae for determining the distance and recessional velocities of extremely remote galaxies. This enabled them to compare expansion rates over a much broader time span in the history of the universe. They were now in a position to measure just how much the universe was slowing down over those cosmic time spans.

And that's when the biggest and most unexpected surprise in the history of cosmology hit them. When they analyzed and re-analyzed their data, both teams came to the same startling conclusion. *The universe was not slowing down, it was speeding up.* And it has been doing this for last five billion years or so.

This suggests that some sort of antigravity force, which they call dark energy, is a previously unknown property of space itself. By comparison, the dark matter mystery seem almost unimportant. Why? Because calculations show that this dark energy makes up a whopping 73 percent of the mass and energy of the universe. Dark matter now gets downgraded to 23 percent and ordinary matter (the only stuff we are familiar with) comes in dead last at 4 percent. And so the cosmos gets stranger still.

What We Don't Know

Nobody knows what this dark energy is, even though it makes up 73 percent of the universe. At this point in time, it's even more mysterious than dark matter. And even more embarrassing for our cosmologists. (I hear some of them don't even go to parties anymore.)

What Might Be

Compared to speculations on dark matter, scientists are pretty vague when it comes to dark energy. Confusion as to the nature of dark energy was very evident after its announcement in 1998. Just look at the number of different labels found in the scientific literature to describe this weird phenomenon at that time:

- dark energy
- cosmological constant
- vacuum energy
- vacuum energy density
- energy density of space
- energy of the quantum vacuum
- gravity of the quantum vacuum
- quantum gravity

- repulsive gravity
- repulsive force
- outward pressure
- negative pressure
- bizarre form of matter.

That last one sounds ... like ... really desperate.

Some theorists have suggested that because dark energy seems to be a basic property of space itself and since space is increasing with cosmic expansion, when the universe is twice its present age its dark energy content will be 97 percent. So much for matter. All this suggests that a significant modification may be needed to the theory of general relativity. Or maybe the Big Bang theory. Or both. Such is the challenge of scientific progress.

Currently, there seems to be three speculations on dark energy, none of which are very clear or satisfying to us ordinary folk:

- Dark energy is a basic energy property of space and, as such, increases as space itself increases in volume.

- Dark energy is an *unidentified* energy field that fills space (as opposed to an *identified* one like the electric field, the magnetic field, or the gravitational field).

- There is no dark energy. There is a flaw in the theory of general relativity that, if fixed, would explain the changing rate of cosmic expansion over the history of the universe.

Chapter 16 —
What Caused the Big Bang … then Life?

As mentioned back in Chapter 12, some cosmologists chicken out here. They duck this fundamental question about the origin of the Big Bang, arguing that it is meaningless since there was nothing prior to the Big Bang. Not even time. "What caused the Big Bang" is essentially a *how* question that often evolves into philosophical *why* questions. Like "why was there a Big Bang", "why is there a universe", "why is there life". Let's face it, science is pretty good at answering a lot of the "how" questions, but not the "why" questions.

The Big Bang

So what caused the Big Bang? It's only natural to ask these sort of how questions because, in the macro world, we are all familiar with cause and effect. Every event must have a cause, right? The pot of soup got hot *because* someone turned the burner on underneath it. The baseball pitcher just gave up a ninth inning walk-off home run *because* his arm got tired or maybe *because* the batter guessed right on the type of pitch and made a really good swing. The apple fell from the tree to the ground *because* of gravity. The Big Bang occurred *because* … ah … because …

Surely, there is an answer. Events just don't happen for no reason. Not even the Big Bang event. Or so the thinking goes.

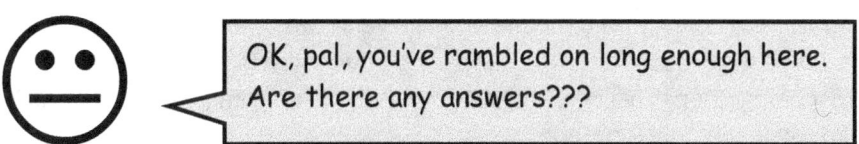

OK, pal, you've rambled on long enough here. Are there any answers???

There are no convincing answers, that's for sure. Only possibilities (or general hypotheses). What follows here is a brief summary of possibilities that seem likely to some scientists. Maybe our readers will have some ideas of their own. We'll call each of these possibilities an **hypothesis** in this discussion. It seems very unlikely that any of these hypotheses would ever obtain the status of a scientific theory (recall Chapter 8).

A Big Bang from a Big Crunch Within Our Universe

With the recent discovery about the expansion of the universe speeding up (see Chapter 15), this first hypothesis for the Big Bang is now looking a little more unlikely. It depends upon the cosmic gravity in our universe being strong enough to eventually bring the cosmic expansion to a halt and start a contraction of all that matter. This is like throwing a ball high into the air and having Earth's gravity eventually bring its upward flight to a halt, then pulling it back down to Earth.

As the cosmic matter gets pulled closer together, its contraction would speed up even more — just like the ball speeds up as it gets closer to the ground. Eventually all that matter would come together at some point in a gigantic collision. Cosmologists call such a collision the **Big Crunch**. The idea is that such a collision would generate enough energy to create another Big Bang and the whole cycle of expansion and contraction would repeat itself. According to this model of our universe, it could have already gone through a number of such cycles — perhaps an infinite number. Or we could currently be in cycle number 14 or 3,744,991, or whatever.

Obviously, we can't go back in time to observe these earlier cycles, so this really isn't a question that science can answer. We can only speculate.

Well, that would explain the latest Big Bang 13.7 billion years ago and all the others that preceded it ... except for the very first one — the one that started the whole series of cycles.

Well, you got that right, Tri-face. This hypothesis says that there has been a number of versions of our universe in the past and that we are simply living in the present version. It doesn't answer the question as to how the first one got started. Or whether each rebirth produces the same sort of universe as the previous one. We have no way of knowing what the previous universe was like. But if true, this hypothesis would mean that there was *something* prior to the latest Big Bang — that time did exist before then. This whole cyclic scenario is like setting the cosmic clock back to zero with each new Big Bang.

This cyclic existence of our universe could have been going on for trillions and trillions of years. Maybe it has always been going on. Maybe there was no beginning.

A Big Bang from a Big Collision Outside Our Universe

This second hypothesis introduces the idea of more than one universe existing at the same time. Rather than multiple versions of one universe separated by time (as described with the Big Crunch), there might be multiple universes co-existing in time, but separated by "space". Chapter 18 discusses this possibility further and calls it the **multiverse**.

In this hypothesis, our Big Bang was caused by the energy produced from a collision between two of these universes. There could be many such collisions going on all the time in this multiverse, creating new universes all over the place. We can think of no way that we would ever be able to observe any of these other universes, so this really isn't a question that science can answer. We can only speculate. Sound familiar?

As with the Big Crunch hypothesis, a multiverse clearly means that there was something prior to our Big Bang. But, again, as Tri-face is probably dying to point out, this explanation only covers our Big Bang. It doesn't explain how such a multiverse began.

A Big Bang from ... well ... Nothing

If you think the Big Crunch is hard to imagine, or a collision between two universes is hard to imagine, wait until you read about this third hypothesis. In this one, our universe pops into existence from nothing. It just happens.

An effect without a cause!!! Have cosmologists gone crazy?

No, but they have been influenced by the particle physicists who deal with quantum mechanics. We have mentioned only one or two characteristics of quantum mechanics in this book, but now it's time to mention another one. Quantum events occur in the micro world at the atomic level, so we don't see these events in our everyday lives. We don't appreciate just how weird the quantum world is.

The startling fact is that the mechanism of cause and effect doesn't always apply in the quantum world. Things really do seem to happen for no reason. Events involving atomic particles can occur quite randomly and scientists have no way of predicting them. Individual particles can simply pop into and out of existence at any time. Particle physicists have actually observed this sort of weirdness in particle accelerators and can demonstrate it experimentally.

Looking at this weirdness another way, scientists can determine the chances (or probability) of certain events occurring within a given timeframe, but they cannot predict which particles will take part in these events, nor when.

Radioactive materials provide a good example of this strange concept. The atomic nuclei of these materials experience radioactive decay — that is, they break down spontaneously into the nucleus of a different chemical element with the release of energy. This means that at any unpredictable time, any one particular nucleus of the material could suddenly decay.

Scientists have determined the rate at which this radioactive decay occurs. For a given material, they know how long it takes for the decay of half of the material — a time period known as the **half-life** of that material. Half-life values range from a fraction of a second to thousands of years, depending on the radioactive material. But — and here's the point — no one can predict which particular nucleus will decay, or when. There is no rhyme or reason why one particular nucleus should decay now and another particular one decay 3.265 seconds later.

In other words, there seems to be no cause and effect at the atomic level. It's just a chaotic world of uncertainty for individual particles. Scientists can only determine the overall probability of behavior, such as the half-life of a given radioactive material.

> I think I get it. Unpredictability at the atomic level. Pretty weird all right. But what does any of this have to do with the Big Bang???

Can't blame you for asking that one, Tri-face. It's all a bit of a stretch. Some cosmologists are suggesting that if our universe started out at the Big Bang moment as an infinitely dense singularity (see Chapter 12), then perhaps this singularity — being so small — behaved just like any unpredictable atomic particle would in the micro world. Perhaps it just popped into existence. Just like those particles that have been observed popping into and out of existence today in science labs. Maybe the effect — namely this cosmic singularity popping into existence — didn't need a cause. Didn't need a Big Crunch. Didn't need colliding universes. Maybe it was just one of those quantum events. It just happened.

So here's a really philosophical question for you. You don't have to answer it right away. Take a few years to think about it.

Which do you, personally, find easier to imagine?

A cyclic universe or a multiverse that ...

(1) somehow began from nothing

OR

(2) had no beginning, but rather, has always existed.

Life

We seem even further in the dark when we ask about life. If there is anything more awesome than the existence of a universe, it's got to be the emergence of life within previously lifeless surroundings. Evolutionary theory gives us a pretty good idea how life evolved on this planet from **microbes** (mostly single celled life forms) all the way up to the complex life forms that populate the globe today. But nobody knows how life began. How did we initially get from lifeless lumps of matter to those first living cells? Today's theorizing and laboratory experimentation on the subject have left scientists not even knowing if they are close to the answer or not.

Astrobiology

There is a whole field of science called **astrobiology** that is working towards an answer to this perplexing question, plus other questions about life beyond our planet (the subject of the next chapter). The guys and gals working in this field are known as astrobiologists. They have examined some really ancient fossils from living cells and claim that life arose on Earth roughly 3.5 billion years ago. It may have risen even earlier, but our planet was heavily bombarded with large asteroids and comets up to 3.8 billion years ago. So astrobiologists figure that any earlier appearances of life would have been repeatedly wiped out during this bombardment. Life may have originated and disappeared several times before Earth's environment became calm enough for it to gain a permanent foothold.

There was no ozone layer in the atmosphere back then to protect living organisms from harmful ultraviolet (UV) radiation. Therefore, it is most likely that life originated either below the planet's surface or underwater beside hot hydrothermal vents on the ocean floor. And — as everybody knows — life on Earth is carbon based. It is made up of complex molecules formed from a few common elements, particularly carbon. The most abundant elements in your body, in order by weight, are oxygen, carbon, hydrogen, nitrogen, calcium, and phosphorus.

Scientists speculate that there were three sources of simple organic molecules in Earth's early days — the atmosphere, the hydrothermal vents in the oceans, and the meteorites that bombarded the planet. These simple organic molecules would be available as input to the chemical reactions that produced the more complex organic molecules that form the basic building blocks of life.

 NOTE: Don't worry, we're not going to subject anyone to a crash course in organic chemistry. Suffice it to say that the simple molecules in the sources mentioned above include ones such as nitrogen (N_2), carbon dioxide (CO_2), ammonia (NH_3), methane (CH_4), and water (H_2O). The more complex molecules (the basic building blocks of life) that were produced from these simple ones included proteins, enzymes and amino acids, among others.

A Recipe with Ingredients but No Cooking Instructions

All this chemistry that was present in the early oceans was like a **primordial soup** — the ingredients for forming living organisms. However, the mystery of how these ingredients evolved through further chemical processes to actually form living organisms remains a mystery.

Immigrants from the Red Planet, Mars?

Finally, astrobiologists do not rule out the possibility that life did not originate on Earth. It may have arrived here as living microbes in a dormant condition buried safely within meteorites or asteroids that dropped in from outer space. Scientists have determined that some of the meteorites found on Earth came from Mars.

 NOTE: Bubbles of gas trapped within these meteorites have been analyzed and found to match the composition of the Martian atmosphere. Non-manned explorations to the red planet have provided us with the necessary data to make such a comparison.

Just like Earth, Mars was bombarded with large objects from space a few billion years ago. Inevitably, these collisions would dislodge large chunks of the Martian surface into space at a speed greater than the escape velocity of that planet. After millions of years orbiting Mars or other planets farther afield (including Earth), some of these chunks of debris would eventually land on Earth. If some of them contained living microorganisms, scientists have reason to think that these organisms could remain dormant and survive the long journey in outer space. If so, they could well have been our source of life on Earth.

 You mean we could all be descendents of microscopic Martians??? How depressing.

Not depressing at all, Tri-face. It's a fascinating possibility. Aside from Earth, Mars is the most likely planet in the solar system to give birth to life. And being smaller than Earth, it would have cooled down enough to permit such a birth long before Earth did. Perhaps you might feel better about all this if I point out that the opposite might have happened instead. Microbic life from Earth could have ended up on Mars by the same sort of process, in which case any life on Mars would be descended from microbic Earthlings.

This process of life forms being transported in space from one celestial body to another is know as **panspermia** (but you probably won't find that word in any dictionary lying around the house). The concept of panspermia, however, does not answer our original question about the origin of life. It only offers a possible change in the location of that origin. So we still don't have an answer to how life began.

Have you considered a career in astrobiology? We need answers!

Chapter 17 —
Hello, Hello. Anybody Out There?

Ever since mankind has come to understand the size and nature of the universe, people have been asking if there is any intelligent life beyond Earth. Different labels have been used to described such life:

- Extraterrestrial — beyond our world
- Alien — very different from us (like "little green men")
- Complex — lots of complex biological structure, like ourselves
- Intelligent — lots of mental capacity, like ourselves.

Maybe we should be asking if there is any intelligent life on Earth. If aliens landed here today and looked around to see what we creatures are doing to ourselves and our planet, I doubt that they would consider us to be very intelligent.

Can't disagree with you there, Tri-face. We seem to be misusing what little intelligence we do have — to the point of threatening our long term survival as a species. But that's a whole different issue. I think it's safe to say that if aliens did drop in for a visit, they would be much more intelligent and much more advanced than the human race. We humans have only been around for about one million years and have just barely reached the stage of transmitting radio signals and tinkering with a bit of space travel. On the cosmic timescale, we are infants at technology.

Any aliens capable of visiting us would be millions of years ahead of us. Millions! Remember that the Milky Way galaxy has been around and forming stars and planets for roughly ten billion years.

One of the key issues about extraterrestrial life is whether or not stars, in general, have planets. After all, you can't expect life to form and evolve on a star — a flaming hot ball of nuclear fusion. You need the much cooler environment of land and water that some planets and their moons may provide. In the 1990s astronomers perfected their observational technology to the point where they could start searching for planets around neighboring stars. They wanted to know if our solar system with its eight planets (sorry Pluto) is a rare thing in the cosmos or commonplace. They labeled planets beyond our own solar system **extrasolar planets**, but soon abbreviated that to **exoplanets**. Astronomers thrive on buzzwords, just like everyone else.

In this chapter, we will first look at how astronomers search for exoplanets. We then present two opposing viewpoints about the probability of extraterrestrial intelligence (abbreviated ETI in this chapter). Take your pick on which viewpoint you think is more likely.

Discovering Exoplanets

There are two big reasons to search for these planets: (1) to gain more insight into the evolution of solar systems, and (2) to move one step closer to finding ETI life in our galaxy.

Unfortunately, it is almost impossible for astronomers to actually see an exoplanet through a telescope. This is because planets orbit extremely close to their parent star (compared to the vast distances between stars) and their reflective light is extremely dim compared to the overwhelming glare of that star. It would be like trying to spot a tiny firefly hovering just above the flames of a raging campfire. But astronomers are a very determined bunch. They have come up with a technique that allows them to detect exoplanets *indirectly*.

A star's powerful gravitational field holds its planets in their orbits. But these planets have a gravitational field of their own and even though it is a lot weaker, it still has some effect on the star. On one side of the planet's orbit it tugs on the star one way and on the opposite side of its orbit it tugs on the star the other way. So as the star moves through space, it actually wobbles back and forth a bit because of these planetary tugs. When astronomers record the spectrograph of such a star over a few days (or years in some cases), they can detect this wobble, thanks to the Doppler effect.

Figure 17-1 illustrates how this is done. For simplicity, let's assume the star has only one planet big enough to cause an observable wobble. As the star's wobble moves *toward* Earth, its light gets blue shifted along our line of sight because of the Doppler effect. (You do remember the Doppler effect, right?) When it moves *away* from Earth, this light gets red shifted. As you learned in Chapter 9, astronomers can measure the amount of these wavelength shifts and from that amount calculate the velocity of the wobble. These velocities are incredibly small in the case of a wobbling star, so the wavelength shifts are also small and therefore very difficult to measure. But astronomers have refined the technique down to the point where they can detect a wobble velocity as low as three meters per second.

As you can see from the diagram, this wobble actually forms a tight little orbit around the center of gravity of the star/planet system. The time it takes to complete one of these stellar-wobble orbits represents one year in the life of the planet. Knowing this time period *and* the really slow velocity of the wobble *and* the mass of the star, astronomers can calculate the unseen planet's orbital velocity, its mass, and its distance from the star. All this from a slight wobble many light-years away. Science in action!

 NOTE: These calculations, however, usually understate the exoplanet's mass and orbital velocity because the orbit of the star's wobble will usually be at some angle to our line of sight rather than edge on. Because of this we are measuring a wobble velocity that is actually less than its true velocity. The plane of the wobble would have to be edge on with our line of sight to get a true value. So when astronomers report a discovery of an exoplanet with a calculated mass of two Jupiters, they say that the new found planet has a mass of *at least* two Jupiters.

These mass figures are typically reported in Jupiter masses because most of the exoplanets are about the size of Jupiter or bigger. A few smaller planets have been found, but they are still several times larger than Earth.

This uncertainty about the actual mass and orbital velocity of an exoplanet is just one drawback of the method. Another drawback is that the less massive or farther out an exoplanet is, the less wobble it causes on the parent star. As a result, Earth-sized planets orbiting at a comfortable distance from their star (planets potentially suitable for life) are almost impossible to detect. This, of course, doesn't mean that no such planets exist.

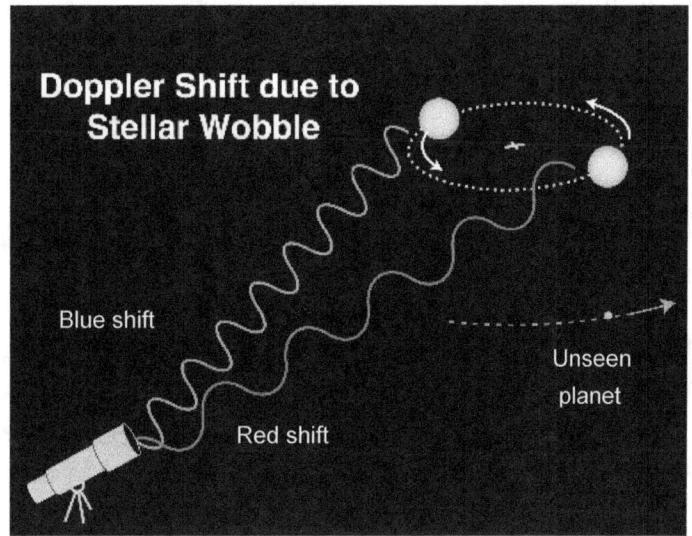

Figure 17-1 Detecting an Exoplanet from Its Sun's Wobble

As of March, 2013, close to 900 exoplanets have been found. Most of them, however, are not only Jupiter size or larger, they orbit very close to their star where it is too hot for life. Some of them are so close they actually complete an orbit in a matter of days. Astronomers, being a very imaginative bunch, call these exoplanets **hot Jupiters**. Such solar systems are not like ours. Our system has small terrestrial planets at moderate distances from the sun and large gas planets like Jupiter much farther out. Does this mean our type of solar system is rare or just plain hard to find? Astronomers aren't yet sure.

The SETI Viewpoint

SETI stands for Search for ExtraTerrestrial Intelligence. The SETI Institute has been searching for meaningful radio signals from outer space since the 1980's. By "meaningful" we mean signals that clearly come from an intelligent source. The astronomers at SETI haven't detected any such signals yet, but they are obviously of the opinion that there is intelligent life out there somewhere. Their argument is based on statistics — there are many billions of stars in our Milky Way galaxy alone, so there are probably many billions of planets. It seems ridiculous to think that we, on our planet Earth, would be the only form of intelligent life in the whole galaxy, let alone the universe. You would expect that a large galaxy like ours would either have no life within it or lots of life within it, but certainly not just one isolated case of life (us!) on one planet. This one-case scenario for life seems ridiculously unlikely.

The Drake Equation

Frank Drake, the American astronomer who founded SETI, triggered debates on this subject by developing a general formula for the probability of ETI elsewhere in our galaxy. It became known as the Drake Equation. It included the basic factors that go into a calculation of N — the number of intelligent civilizations in our galaxy that are currently transmitting radio signals that are strong enough for other galactic civilizations (such as ours) to receive. To get this number, you simply multiply all the factors together. Drake reasoned that if the calculation produced a big enough number, it would be worth the effort for astronomers to search for those radio signals.

Unfortunately, nobody knows what values to assign to most of these factors. While some of them can be estimated, based on what we know about stars and planets in our galaxy, others are complete guess work. There are a few variations on the Drake equation, because different factors can be combined to reach the same end. The equation below is one variation. Following this equation, we give an explanation of each of its factors (or terms). And — just to illustrate the use of the equation and have some fun with it — we're going to assign a value (a guess in many cases) for each term and see what we get for N.

You might want to assign different values and see what you get.

Just as a reminder, this equation addresses only civilizations in our galaxy that are communicating over interstellar distances via radio signals:

$$N = N_s \, f_s \, f_p \, n_z \, f_l \, f_i \, f_c \, f_L$$

where:

N = number of civilizations communicating over interstellar distances
Ns = number of stars in our galaxy
fs = fraction of these stars that are sun-like
fp = fraction of these sun-like stars that have planets (solar systems)
nz = number of planets within the habitable zone of these solar systems
fl = fraction of these habitable zone planets where life arises
fi = fraction of these life producing planets where ETI evolves
fc = fraction of these ETI planets communicating over interstellar distances
fL = fraction of the star's lifetime in which this communication takes place.

We used a figure of 400 billion for Ns back in Chapter 1, so we'll stick with that. This may be the most reliable value in the whole equation.

Sun-like stars (fs) means those stars that are roughly the same mass, age, and metallicity as our sun. This fs fraction is needed here because stars that are much bigger than our sun do not live long enough to allow any potential life on its planets to evolve all the way to ETI. (Recall in Chapter 13 how massive stars live fast and die young.) Stars much less massive than our sun may not provide enough energy to allow complex life to evolve. Stars much younger than our sun have not been around long enough to allow complex life to evolve on their planets. Finally, our sun (and therefore its inner planets) is a metal-rich star.

NOTE: You may remember from Chapter 2 that when astronomers use the word "metal" they mean all the chemical elements except hydrogen and helium. So we have to humor these folks here and include elements like carbon and oxygen, as well as heavy ones like iron and uranium, under the label "metals".

Metallicity is very important because heavy metals are needed to support complex life. Hydrogen and helium are not enough. Stars located near the outer edge of the galaxy are not metal-rich. Stars located near the galactic center are too metal-rich and live in a stellar-crowded violent region not suitable for the evolution of complex life. We'll set the fs fraction to 0.1 (thereby assuming ten percent of all stars in the galaxy are sun-like).

When Drake founded SETI, astronomers had no idea what value to use for the fp fraction. Since then, however, they have made great strides on this issue, as already covered in this chapter. It now appears that planet formation is quite common during the birth of stars, so we'll set the fp fraction to 0.5 (thereby assuming 50 percent of all sun-like stars have planets).

The habitable zone of a solar system is that range of planetary distances from the star that is just right for life. Not too hot, not too cold. The sort of temperatures where lots of water can exist in liquid form. In our solar system, for example, we're sitting on the only planet that lies comfortably within the habitable zone. Venus is one planet closer to the sun than us and has a temperature way above the boiling point of water. Mars is one planet farther from the sun than us and has a temperature below the freezing point of water. Lucky Earth. Lucky us. Let's set the number nz to 1, thereby assuming that, on average, stars with solar systems have one planet in their habitable zone (just like in our solar system).

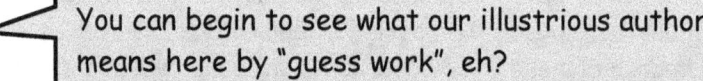

You can begin to see what our illustrious author means here by "guess work", eh?

Oh, for sure. The whole point of Drake's equation was to get astronomers thinking about this topic and promoting the idea of using radio telescopes to search for signals from aliens.

Now the guess work really gets wild. As we saw in the last chapter, nobody knows how life began on Earth, so nobody knows the probability of life starting up on any of the habitable zone planets. Astronomers like Frank Drake and Carl Sagan tended to be very optimistic about this. They believed that if life sprang into existence very early in Earth's history, it would naturally do so on any planet within a habitable zone. So they set the fl fraction to 1 (thereby assuming 100 percent of all planets in the habitable zone spawn life). This seems awfully optimistic, but we'll do the same. Who are we to disagree with these geeks?

This next factor, fi, is one of the most controversial ones of all. What are the odds of getting from microbic life forms — the basic beginnings of life — to complex life with enough intelligence to build radio telescopes??? This long evolutionary process needs a stable environment over eons of time. No sudden and radical changes to climate. No mass extinctions of life due to a collision with a massive celestial object like a huge asteroid, comet, or even another planet or moon. The SETI folks get really optimistic again and set the fi fraction to 1. They figure that's what Charles Darwin would do because intelligence has great survival value. Survival of the fittest and all that. As you will see in the alternative viewpoint later in this chapter, not everyone agrees. We'll avoid taking sides here by setting fi to 0.5 (thereby assuming 50 percent of all planets that spawn life enable that life to evolve into ETI).

Not every intelligent civilization may want to communicate across interstellar space. There could be lots of reasons for this:
- Some of them may be too busy trying to survive.
- Some may be too busy pursuing other interests like hockey or crossword puzzles or modeling the origin of the universe.
- Some may feel it's just a waste of time.

- Some may simply be antisocial and have no interest in saying "hello" to their neighbors.
- Some may be too scared to say "hello" for fear it will reveal their location to a hostile neighbor who might then invade their planet.
- Some may be so advanced compared to us that they are communicating in some way other than radio signals — some way that we don't even know about.

So we will set the fc fraction to 0.3 (thereby assuming 30 percent of intelligent civilizations are communicating over interstellar distances via radio signals). Why this particular fraction? Because we're tired of using 0.1 or 0.5 or 1 and, like everybody else, we don't have clue what a reasonable value would be.

Last, but not least in importance, is the fL fraction. No civilization transmits radio signals for the entire lifetime of its sun. Any existing civilization would take a very long time to evolve and would presumably die out some time later. So we have to include a time limit in the Drake equation that reflects this. Although civilizations will die out when their sun runs out of nuclear fuel (recall Chapter 14), they are far more likely to die out much sooner from some astronomical-type mass extinction. Either that, or some self inflicted disaster, such as nuclear war, over population, mismanagement of the planet's resources, famine, pollution, disease, or runaway global warming.

 NOTE: You can probably think of a few more possibilities. Not to get too depressed about this, but how much longer do you think mankind will last on our planet the way things are going?

We factor in this time limit by dividing the duration of interstellar communication from an ETI civilization by the lifetime of its sun. This gives us the fraction of that sun's lifetime during which one of its planets is broadcasting ETI communications. So, let's assume the average ETI civilization communicates for 10,000 years before it dies out or gets tired of sending out signals. And let's assume that its sun-like star has an expected lifetime of roughly ten billion years. That gives us an fL fraction of $10,000 / 10,000,000,000 = 0.000001$ (or 10^{-6}).

OK, now let's have a drum roll while we calculate N from the Drake equation using the values selected above ...TA DA!

$$N = Ns\ fs\ fp\ ne\ fl\ fi\ fc\ fL$$

$$= (4 \times 10^{11}) \times 0.1 \times 0.5 \times 1 \times 1 \times 0.5 \times 0.3 \times 10^{-6} = 3,000$$

Three thousand doesn't seem like a lot of civilizations from a galaxy of 400 billion stars, but that's our answer and we're sticking to it. Tri-face agrees (for whatever that's worth). When this issue was debated in the 1980s, Frank Drake came up with 10,000 and Carl Sagan estimated a cool one million.

While this whole statistical approach suggests at least a few thousand intelligent civilizations in our galaxy alone, there is another viewpoint. In 1950 a discussion on ETI took place among atomic physicists at the Los Alamos National Labs. When it was argued that there must be a lot of alien colonization in our galaxy, the renowned Italian nuclear physicist, Enrico Fermi, looked around the room and casually asked, "So, where is everybody?"

In other words, why haven't we seen them? These aliens, if they exist, would have been around for millions and millions of years and probably traveled all over the galaxy. They would have surely visited Earth by now. So why do we see no sign of them having been here? And why do see no signals from them from outer space?

Could it be because they don't exist?

The Rare Earth Viewpoint

At the turn of the century, a couple of American scientists, Peter Ward and Donald Brownlee from the University of Washington, took a different approach to this whole issue. In their 2000 book, *Rare Earth*, they questioned some of the values that Frank Drake and Carl Sagan plugged into the Drake equation. But more importantly, they used their combined knowledge of geology, mass extinctions, astrobiology, solar system origins, comets, and meteorites to examine planet Earth as a suitable place where complex life could evolve. Their conclusion was that while microbic life might be commonplace in the universe, complex life is probably not. They argued that the conditions necessary for the evolution and survival of higher life forms are so complex and fragile that they are unlikely to be available in many places other than Earth.

WOW! That's quite a conclusion. They're saying ... like ... Earth is a very rare planet. Are they suggesting that we humans may be alone in the universe? That we may represent the only intelligence there is???

They don't' go so far as to suggest that we are alone. But they obviously believe ETI is very rare in the universe. If they're right, it becomes even more important that we take care of this precious planet that we live on — the one that was so dramatically photographed on one of the Apollo missions to the moon a few decades ago. See Figure 17-2.

Figure 17-2 Earth as Seen Above the Moon's Limb

This so-called **Rare Earth Hypothesis** argues that (1) there is a long list of favorable conditions on our planet that made it possible for life to evolve to its present level and (2) the odds are very low that all (or even most of) these conditions would be available elsewhere. In the remainder of this chapter, we look at some of these conditions to give you an idea of what Ward and Brownlee mean by "rare Earth".

 NOTE: Bear in mind that this hypothesis is based on just one sample of life — the one here on Earth. It therefore assumes that a carbon-water-based life is the only kind possible. For reasons of chemistry, it rules out little green men made out of, say, silicon as being extremely unlikely.

Habitable Zone Gives Us a Moderate Climate

As mentioned during our discussion on the Drake equation, Earth is right in the middle of the habitable zone of our solar system. This gives life a couple of its fundamental requirements — moderate temperatures and a vast reservoir of liquid water in our oceans and lakes. If you compare the freezing and boiling points of water (zero and 100° C) with the planetary temperatures in Table 5-1 on page 36, you will see that the habitable zone is quite narrow.

But it's really narrower than Drake and Sagan assumed, because life needs a few billion years to evolve from microbic to complex. During all that time the planet must remain within the habitable zone. Unfortunately, stars grow a lot hotter during their lifetime. So the inner boundary of the habitable zone continually gets moved farther out from the star with time, making the width of this zone narrower over the billions of years that life is evolving. Only a very small percentage of planets would remain within this continuously narrowing habitable zone long enough to allow ETI to emerge from the evolutionary process. A much smaller percentage than Drake and Sagan assumed. Lucky us. Earth is one of the "very small percent". Without this comfortable position, we probably wouldn't be here today.

Big Brother Jupiter Protects Us from Space Junk

As the largest planet in our solar system, Jupiter has done a great job of devouring much of the debris left over from the formation of the solar system. It's location and gravitational influence has been very helpful in keeping Earth's neighborhood relatively free of dangerously large celestial bodies. It is estimated that the current impact rate on Earth for objects ten kilometers or more is one collision every 100 million years. The last one was 65 million years ago and it wiped out the dinosaurs. Without Jupiter, it is estimated that this impact rate would be about 10,000 times higher. This works out to one impact every 10,000 years. Imagine the damage that sort of impact rate would have done to the evolutionary process on Earth.

I wonder what percentage of planets within the habitable zone of their sun have a big brother like our Jupiter standing nearby. Lucky us. Without this protection, we probably wouldn't be here today.

Big Moon Protects Us from Going Tipsy

It would be very difficult for life to evolve from microbic to complex if a planet's climate suddenly changed completely every so many millions of years. But that's what happens to most planets. The tilt of their axis (what astronomers refer to as **obliquity**) suddenly gets drastically altered because of the gravitational pull of their sun and other large planets when they line up in certain ways. And — as everybody knows — it is this angle of tilt that determines the seasons and provides a stable climate for each hemisphere. Figure 17-3 shows Earth's obliquity as 23.4 degrees.

This stable climate gets thrown out of whack when a planet's obliquity gets suddenly changed. Scientists have reason to believe that, except for Earth, the tilts of the terrestrial planets in our solar system have chaotically flipped all over the place throughout their lifetime. Mar's obliquity is about the same as Earth's right now, but in the past it has varied by 45 degrees or more.

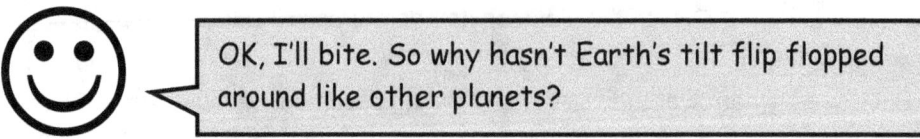

OK, I'll bite. So why hasn't Earth's tilt flip flopped around like other planets?

Thought you'd never ask. It's because Earth has a massive moon with enough gravitational presence to mask out those annoying gravitational pulls that throw other planets out of whack. Our moon holds Earth's tilt in a very narrow range — it varies from 22.1 to 24.5 degrees over millions of years. This is a big reason why Earth has had a relatively moderate and stable climate for so long. Without our moon, scientists figure that Earth's obliquity would flop around in a range from zero to 85 degrees. The moon's gravitational effect has also slowed down Earth's daily rotation and provided ocean tides, both of which are beneficial to the evolution of life.

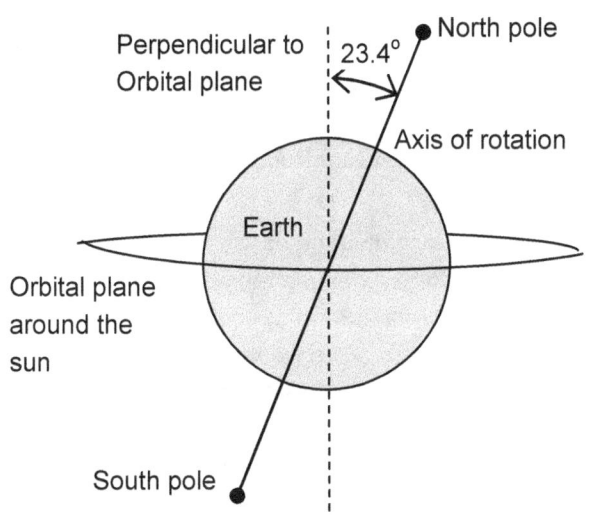

Figure 17-3 Earth's 23.4°

Well, that's all pretty convenient. Too bad every planet doesn't have a massive moon like ours. So where did we get this moon from anyway?

Well, I guess you could say we got lucky again. The most likely theory about the moon's origin is the collision theory. When Earth was very young and still molten hot, it was hit at a glancing angle by a massive body perhaps as big as Mars. The resulting collision spewed a massive amount of matter from both the intruding object and Earth's surface out into space. This matter stuck together to form the moon, which went into orbit around Earth. Since then, the orbit has slowly increased its distance from Earth and continues to do so today.

Samples of the moon's surface brought back by the Apollo missions have added much support to this collision theory. Analysis of these samples showed them to be remarkably similar to the chemical composition of Earth's crust.

I wonder what the odds are for a collision like that? The result would have been very different if the colliding object had been much smaller or much larger. Or if the angle of collision had been different. Or if it had happened later in Earth's lifetime when it wasn't so molten hot. It seems unlikely that many other planets would be lucky enough to give birth to such a massive moon in this fashion. Lucky us. Without such a moon, we probably wouldn't be here today.

Continental Drift Gives Us Land and a Global Thermostat

You've probably heard of continental drift. It's the movement of Earth's crust in large chunks or **plates** (as geologists call them) over the hot flowing material underneath the crust. Geologists call this material Earth's **mantel**. Today these rock-crazy scientists refer to this process by the fancy label **plate tectonics**. So how does this help the evolution of ETI?

Well, all this movement of Earth's crust — including the ocean floor — scrunches up crust material here and there to form mountain ranges. This, in turn, leaves behind scooped out areas that fill up with water to form oceans and lakes. Just think what we would have without plate tectonics. We would have a water planet — a planet with one big shallow ocean covering every square meter of the globe. No land. Now that would surely limit the extent to which evolution could go. We might get as far as dolphins on the IQ scale, but that would be it. No agriculture. No books. No cities. No cars or TV or cell phones. And certainly no space trips to that precious moon of ours.

 NOTE: Most of the water is thought to come from the icy comets that bombarded Earth so frequently during its early history.

Plate tectonics also gives us a thermostat to control our global temperatures. It works like this. The carbon dioxide (CO_2) that living cells and burning forests produce goes into the atmosphere where it causes global warming. But this warming raises the temperature of the oceans, which, in turn, causes them to absorb more CO_2 from the atmosphere which then lessens the global warming effect. Eventually, the CO_2 in the oceans ends up in Earth's crust beneath the ocean floor as carbonates. Meanwhile, the forceful motion of plate tectonics creates volcanoes that spew hot molten material from the depths of the Earth into the atmosphere, along with some of that CO_2 that was deposited in the Earth's crust as carbonates. This CO_2 has now gone full circle. It is back in the atmosphere where it started. This process is known as the **carbon cycle** and — as long as modern industry doesn't upset this balance too much with its carbon emissions — it will continue to keep our global temperatures under control.

Not all planets have plate tectonics. A planet needs a good source of heat within its interior to keep the mantel material molten hot. For this, we need a planet big enough not to cool down too quickly after its formation. And we need a planet that can maintain its interior temperature by having an additional source of heat. In the case of Earth, this heat source comes from radioactive materials with a long half life — the radioactive forms of "metals" like potassium, thorium, and uranium. Only a metal-rich planet of sufficient size can provide this. Earth is one such planet.

In fact, planetary scientists and astrobiologists tell us that Earth is the only place in the entire solar system that has plate tectonics today. And, of course, the only place where we have detected any kind of life. Lucky us. Without plate tectonic, we probably wouldn't be here today.

Magnetic Field Protects Us from Cosmic Radiation

All that heat and plate movement in Earth's interior causes motion within the planet's molten iron/nickel core which, in turn, creates powerful electrical currents there. As you may have learned in science class, an electrical current produces a magnetic field around it. It is because of these powerful electrical currents in Earth's core that our planet has a strong magnetic field that reaches far out into space.

Yeah, so who needs it?

We do. Without that magnetic field, we would be bombarded by the deadly cosmic radiation that exists in outer space. This so-called "radiation" is actually very energetic, fast moving particles of electrons, protons, and heavier atomic nuclei that, if allowed to reach Earth, would be very harmful to living organisms. It would also gradually strip away our atmosphere. It would prevent complex life from ever getting a foothold on our planet. But because this "radiation" is actually charged particles, Earth's magnetic field is able to deflect most of it away from the planet's surface.

None of the other terrestrial planets in our solar system have a comparable magnetic field. In fact, Venus and Mars have virtually no magnetic field.

Thank goodness for all that heat in Earth's interior, due to its size and all those radioactive heavy metals there. It's that heat that drives plate tectonics and creates the electric currents that provide us with a protective magnetic field. I wonder how many exoplanets are that lucky. Without that heat, we probably wouldn't be here today. Is this beginning to sound like a broken record?

There are a number of other favorable conditions for life on Earth that we have not covered in this chapter (our atmosphere, for example). I wonder how many exoplanets have the right combination of favorable conditions for the evolution of ETI. Maybe the folks promoting the Rare Earth Hypothesis should call it the "Lucky Us Hypothesis."

Chapter 18 —
Why is the Universe So Fine-tuned for Life?

This is certainly not the sort of question that pops up in everyday conversation. What do we mean by "fine-tuned for life" anyway? Well, as it turns out, there are a handful of physical constants in nature that seem to have just the right numerical values for producing the sort of universe we see today — one that is long living and teaming with galaxies, stars, planets, and (in the case of at least one planet) *life*. If these numerical values were just slightly different, our universe would be a very different place and we humans would not be around to see it.

This intriguing concept has been pointed out by a number of cosmologists lately, including Sir Martin Rees, author of *Just Six Numbers*, first published in 1999. In this closing chapter, we will briefly describe just three of these numbers to give you some idea of how lucky we are to have the sort of universe we do.

Electromagnetic Force Verses Gravity

The electromagnetic (EM) force is one of nature's fundamental forces and is responsible for the interactions between electrically charged particles. It behaves as a force of attraction between particles of opposite charge (such as a proton and an electron) and a force of repulsion between particles of like charge (such as two protons), so it plays a major role in the micro world. It holds atoms and molecules together. Electricity and magnetism are consequences of this force.

Gravity (G) is another of nature's fundamental forces, but it is unbelievably weak compared to electromagnetism. In fact, the EM force is 10^{36} times stronger than gravity.

$$\frac{\text{EM force}}{\text{G force}} = 10^{36}$$

This helps to explain why a tiny magnet can cause a metal paperclip to jump up from the top of your desk — even though the gravitational pull of the entire planet Earth is trying to hold it down. Nevertheless, it is gravity that rules the larger macro world. As you have seen in this book, it controls the motions of the planets and galaxies. This is because gravity only interacts with mass as a force of attraction, while electromagnetism interacts as both attraction and repulsion. So the effect of electromagnetism cancels out when sizable quantities of mass are involved because the overall electrical charge of such mass is neutral (roughly the same number of protons as electrons).

Cosmologists tell us that if the G force was just a tiny bit stronger relative to the EM force — so that the EM/G ratio was 10^{30} instead of 10^{36} — things would be quite different:

- It would be a miniature universe because it would take a billion times fewer atoms to make a typical star or planet because of the extra gravity created by those atoms. Instead of lasting for ten billion years (like our sun), such a star would burn out after about ten thousand years.

- Similarly, if any living creatures did manage to appear, they could not grow larger than insects because gravity would have a stronger pull relative to the EM force holding those creatures together.

- The universe would quickly stop expanding and collapse in a Big Crunch.

- None of these effects favor biological evolution of any kind. In such a universe, we humans would not exist.

And all because of a difference of only 10^6 in an EM/G ratio of 10^{36}.

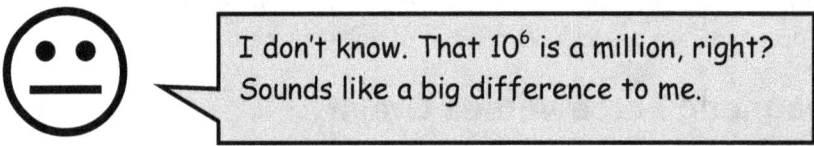
I don't know. That 10^6 is a million, right? Sounds like a big difference to me.

Not when you compare it to 10^{36}. The percentage difference is ridiculously small:

$$\frac{10^6}{10^{36}} \times 100 \text{ (to get a percentage comparison)} = 10^{-28} \%$$

That's 0.0000000000000000000000000001 percent

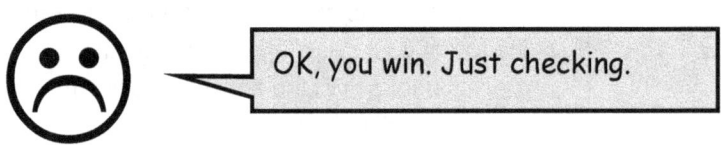
OK, you win. Just checking.

Stellar Nuclear Efficiency

Back in Chapter 13 we learned that 0.7% of the nuclear mass taking part in the fusion of hydrogen into helium gets converted into energy. Think of this 0.7% rate as the nuclear efficiency in the cores of stars during their lifetime. This particular efficiency depends on one of the fundamental forces that we listed in NOTE 1 on page 106. It's a powerful force that works only over very tiny atomic distances, but it holds the nuclei of atoms together despite the protons that exist in these nuclei. This force is called the **strong nuclear force** and it is much stronger than the EM force, which is why the protons don't fly apart in atomic nuclei even though they repel each other electrically.

Well, guess what. This 0.7% number turns out to be most fortunate. Cosmologists and particle physicists tell us that if it had been 0.6 %, this slightly lower strength in the strong nuclear force would have prevented one of the essential steps in the series of reactions that fuse four hydrogen nuclei into one helium nuclei. We would have had a universe made up of hydrogen only — no other elements. And therefore no life of any kind.

If the number had been 0.8%, we would have had another problem. This slightly stronger strong nuclear force would have allowed the fusion of two hydrogen nuclei to take place without the need for a series of other reactions. This more efficient nuclear process would have consumed all the available hydrogen shortly after the Big Bang when nucleosynthesis took place. No hydrogen would have remained. This would have meant no fuel to feed the stars. And of course, no water (good old H_2O). And needless to say, no life.

The conclusion here is that if you want a universe with some complex chemistry that can lead to the creation and evolution of life, you better have a stellar nuclear efficiency range close to 0.7%. Not 0.6% or 0.8%. Fortunately, we have.

Cosmic Expansion

This final example of life-friendly numbers is perhaps the most amazing one of all.

The expansion energy from the Big Bang is causing the universe to fly apart, as first observed by Edwin Hubble. But the gravitational force associated with the mass of the universe is working against this expansion. Gravity is trying to pull everything back together.

Kinda like trying to put Humpty Dumpty together again, right?

Yeah, something like that I suppose. Anyway, the question is — who is going to win this tug of war? Cosmologists have calculated the theoretical **critical density** of matter in the universe that would be required to balance these two opposing forces. It amounts to only a few atoms per cubic meter of space. The question then becomes one of comparing the actual density of matter in the universe with this critical density value.

And so, our geeky friends have defined a ratio called omega (the Greek letter Ω) which is the actual density divided by the critical density:

$$\Omega = \frac{\text{Actual density of matter}}{\text{Critical density of matter}}$$

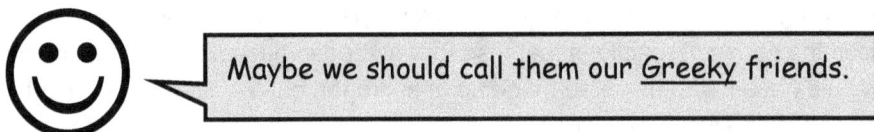

Maybe we should call them our <u>Greeky</u> friends.

Ignoring Tri-face once again, you can see that it is very important for cosmologists to estimate, as best they can, how much actual matter — including the dark matter — there is in the universe. Only then will they be in a position to predict the future of cosmic expansion. The key point is simply that the value of Ω will have one of two results:

1. If the actual density of matter is *less than* the critical density ($\Omega < 1$), gravity will be unable to stop the expansion and the universe will continue to expand forever.

2. If the actual density of matter is *more than* the critical density ($\Omega > 1$), gravity will eventually stop the expansion of the universe, which will then contract and come together in a Big Crunch.

So the fate of the universe turns out to be dependent not only on the two previous numbers we discussed (plus others we have not discussed), but also on the value of Ω. This latter value was initially established immediately after the Big Bang.

Now listen up for a really key point — the Ω value must have been extremely close to a value of one (or unity, as scientists like to say) at that early moment in cosmic history. Why do we say this? Because 13.7 billion years later it is still close to unity at an estimated value of about 0.3. If it had not been extremely close to unity after the Big Bang, that difference from unity would have grown enormously long before now. In the case of result (1) listed above, matter would have expanded too fast for stars and galaxies to have had time to form. Hence no life. In the case of result (2), matter would have stopped expanding and contracted into a Big Crunch long ago. Hence no life.

Surprisingly, we have had neither result so far. We seem to be in that very narrow range that permits lots of star formation and billions of years of time for the creation and evolution of life. It seems our universe was very finely tuned with regard to the initial value of Ω. According to Sir Martin Rees:

"At one second after the Big Bang, Ω cannot have differed from unity by more than one part in a million billion (one in 10^{15}) in order that the universe should now, after ten billion years, be still expanding and with a value of Ω that has certainly not departed wildly from unity."

One part in a million billion is an unimaginable degree of fine tuning. It's kind of like sharpening a pencil and then standing it up on its pointed end without having it fall over.

So … Why is the Universe So Fine-tuned for Life?

Sir Martin Rees outlines three possible responses to this question and I have briefly summarized them below. He personally embraces the third one (Multiverse). Perhaps you can think of a fourth.

1. **Coincidence** — It's just the way it is and we're lucky to be here. Now let's stop wondering about it and just enjoy studying this amazing universe we have. A lot of scientists feel this way. Apparently, it's just too frustratingly philosophical for them to discuss because they feel there is no answer to this "why" question.

2. **Creation by Design** — God did it and that's why it works. And that's why we're here. A lot of folks feel this way. Unfortunately, this doesn't explain where God came from.

3. **Multiverse** — The numbers discussed (in Rees's book) are just far too amazing and fortunate to make the first response believable. And the second response lacks any scientific evidence for a god of any kind. Therefore, it's far more likely that there is another response (or explanation) for the fine tuning — one that involves more than just our universe. There may be a great many other universes, perhaps even an infinite number of them (each created via its own Big Bang). These universes may have evolved differently and therefore have their own set of physical laws and values for the physical constants. The vast majority of them would not be fine-tuned for life, but statistically — with so many of them — some would inevitably be fine-tuned for life. Our universe happens to be one of the fortunate ones.

For some cosmologists, the third explanation is a fundamental argument for the existence of a multiverse. Speculative? Yes. But also quite plausible, since it fits in nicely with the latest ongoing theory in particle physics called **string theory**— a complex and radically different theory from the currently accepted Standard Model of particle physics. But we won't be going there today.

We have now come to the end of this book, leaving you with a lot of unanswered questions. Science has a way of doing that. So does philosophy. When it comes to thoughts of the universe (or multiverse), perhaps the most profound question of all was put forward a few centuries ago by the great German mathematician and philosopher, Gottfried Leibniz, when he asked, "Why is there anything rather than nothing?"

If you have the answer to that, please let us know. Tri-face and I would be interested.

Reference Data

Formulas, Quantities, and Constants

These two pages list some basic formulas and data used throughout the book. They will come in handy if you want to experiment with some calculations of your own.

Circles and Spheres

In the formulas below, R is radius (half the diameter) and π = 3.1416.

Circumference of a circle:	$C = 2\pi R$
Area of a circle:	$A = \pi R^2$
Surface area of a sphere:	$A = 4\pi R^2$
Volume of a sphere:	$V = (4/3)\pi R^3$

Metric System (in brief)

1 kilometer (km)	= 1,000 meters (m)	[0.6 miles]
1 meter (m)	= 10 decimeters (dm)	[3.3 feet]
1 decimeter (dm)	= 10 centimeters (cm)	[3.9 inches]
1 centimeter (cm)	= 10 millimeters (mm)	[0.4 inches]
1 meter (m)	= 1,000 millimeters (mm)	[3.3 feet]
1 metric ton (t)	= 1,000 kilograms (kg)	[1.1 short tons]
1 kilogram (kg)	= 1,000 grams (g)	[2.2 pounds]

Cosmic Distances

1 astronomical unit (AU)*	= 1.496×10^8 km
1 light-year (ly)**	= 9.461×10^{12} km
1 parsec (pc)	= 3.262 ly = 3.086×10^{13} km
1 Mpc	= 10^6 pc = 3.086×10^{19} km

Sun

1 solar mass (M$\odot$)	= 1.989×10^{30} kg
1 solar radius (R$\odot$)	= 6.960×10^5 km
1 solar luminosity (L$\odot$)	= 3.09×10^{26} W (or J/s)

Earth

Mass	= 5.974×10^{24} kg
Radius	= 6,378 km
Escape velocity	= 11.2 km/s

* Average distance from the sun to Earth.
** Distance light travels in one year.

Physical Constants

Speed of light (c) $= 2.998 \times 10^5$ km/s (usually rounded to 3.00×10^5)

Gravitational constant (G) $= 6.673 \times 10^{-11}$ Nm^2/kg^2

Planck's constant (h) $= 6.626 \times 10^{-34}$ Js

Hubble constant (H_0) $= 74$ km/s/Mpc (currently estimated value)

About the Author

Ken Dixon is enjoying his twilight years. He plays tennis to keep his body from turning to slush and dabbles in chess, backgammon, astronomy, and writing to keep his mind from turning to mush. Occasionally he worries about such things as the dismal state of science education in North America, something that prompted him to write *Adventures in the Science of Cosmology* for young teens.

If this book stimulates a few young minds toward a career in science, it will have served its purpose.

Ken earned a BASc in metallurgical engineering from the University of Toronto. In his retirement years, he pursued a life long interest in astronomy by earning a Masters degree in that subject from the Swinburne University of Technology in Melbourne Australia (heck, one can't play tennis all the time).

After a few years of metallurgical research in England, Ken settled down in Toronto for a lengthy career of programming and systems design in the computer industry. A portion of that career included technical writing, teaching, and developing educational material — all useful skills for authoring an educational book on science.

www.ingramcontent.com/pod-product-compliance
Lightning Source LLC
Chambersburg PA
CBHW081448170526
45166CB00008B/2352